OPTICAL COHERENCE TOMOGRAPHY
in Current **GLAUCOMA** *Practice*
Pearls and Pitfalls

OPTICAL COHERENCE TOMOGRAPHY
in *Current* GLAUCOMA *Practice*
Pearls and Pitfalls

Digvijay Singh MD
Ajay Sharma M Optom
Dewang Angmo MD FRCS
Tanuj Dada MD MNAMS

Dr Rajendra Prasad Centre for Ophthalmic Sciences
All India Institute of Medical Sciences
New Delhi, India

JAYPEE BROTHERS MEDICAL PUBLISHERS (P) LTD
New Delhi • London • Philadelphia • Panama

 Jaypee Brothers Medical Publishers (P) Ltd.

Headquarters
Jaypee Brothers Medical Publishers (P) Ltd.
4838/24, Ansari Road, Daryaganj
New Delhi 110 002, India
Phone: +91-11-43574357
Fax: +91-11-43574314
E-mail: jaypee@jaypeebrothers.com

Overseas Offices

J.P. Medical Ltd.
83, Victoria Street, London
SW1H 0HW (UK)
Phone: +44-2031708910
Fax: +02-03-0086180
E-mail: info@jpmedpub.com

Jaypee-Highlights Medical Publishers Inc.
City of Knowledge, Bld. 237, Clayton
Panama City, Panama
Phone: +1 507-301-0496
Fax: +1 507-301-0499
E-mail: cservice@jphmedical.com

Jaypee Medical Inc.
The Bourse
111, South Independence Mall East
Suite 835, Philadelphia, PA 19106, USA
Phone: +1 267-519-9789
E-mail: jpmed.us@gmail.com

Jaypee Brothers Medical Publishers (P) Ltd.
17/1-B, Babar Road, Block-B, Shaymali
Mohammadpur, Dhaka-1207
Bangladesh
Mobile: +08801912003485
E-mail: jaypeedhaka@gmail.com

Jaypee Brothers Medical Publishers (P) Ltd.
Bhotahity, Kathmandu, Nepal
Phone: +977-9741283608
E-mail: kathmandu@jaypeebrothers.com

Website: www.jaypeebrothers.com
Website: www.jaypeedigital.com

© 2014, Jaypee Brothers Medical Publishers

The views and opinions expressed in this book are solely those of the original contributor(s)/author(s) and do not necessarily represent those of editor(s) of the book.

All rights reserved. No part of this publication may be reproduced, stored or transmitted in any form or by any means, electronic, mechanical, photocopying, recording or otherwise, without the prior permission in writing of the publishers.

All brand names and product names used in this book are trade names, service marks, trademarks or registered trademarks of their respective owners. The publisher is not associated with any product or vendor mentioned in this book.

Medical knowledge and practice change constantly. This book is designed to provide accurate, authoritative information about the subject matter in question. However, readers are advised to check the most current information available on procedures included and check information from the manufacturer of each product to be administered, to verify the recommended dose, formula, method and duration of administration, adverse effects and contraindications. It is the responsibility of the practitioner to take all appropriate safety precautions. Neither the publisher nor the author(s)/editor(s) assume any liability for any injury and/or damage to persons or property arising from or related to use of material in this book.

This book is sold on the understanding that the publisher is not engaged in providing professional medical services. If such advice or services are required, the services of a competent medical professional should be sought.

Every effort has been made where necessary to contact holders of copyright to obtain permission to reproduce copyright material. If any have been inadvertently overlooked, the publisher will be pleased to make the necessary arrangements at the first opportunity.

Inquiries for bulk sales may be solicited at: jaypee@jaypeebrothers.com

Optical Coherence Tomography in Current Glaucoma Practice: Pearls and Pitfalls

First Edition: **2014**

ISBN: 978-93-5152-188-4

Printed at: Samrat Offset Pvt. Ltd.

Dedicated to
Our dear alma mater
"RP Centre"

Preface

Glaucoma is the number one cause of irreversible blindness worldwide. Although glaucoma was traditionally diagnosed by a high IOP, optic disc and visual field changes, it was observed that changes in the retinal nerve fiber layer (RNFL) and optic nerve head (ONH) precede visual field damage and the current concept of glaucoma requires the documentation of progressive structural damage to the optic nerve head. Spectral domain optical coherence tomography (OCT) has become a popular investigation for evaluating glaucoma patients as it allows a high resolution imaging of the retinal layers and the optic nerve head along with anterior segment imaging, all of which are critical elements in the diagnosis and follow-up of glaucoma patients. Although easy to perform, the OCT requires a learning curve on performing a high quality scan and it gives a lot of information which is often difficult to interpret for the busy practitioner. Currently, I am facing an alarming situation where patients are being diagnosed with "Red" disease (OCT parameters outside the limits of the normative database are flagged in red color) and started on therapy only based on the OCT when actually they do not have glaucoma. These investigations should only be used to aid the ophthalmologist after a thorough clinical evaluation of the retinal nerve fiber layer and optic nerve head have been performed and should not be used as stand-alone diagnostic tests. Another aspect is the patient's anxiety caused by looking at his/her OCT and getting alarmed by an abnormal result, which actually may only be an artifact or due to a poor quality scan or a large myopic disc.

The practical manual serves as a guide to the judicious use of the OCT for the diagnosis of glaucoma looking at each aspect of OCT imaging—RNFL, optic nerve head, macular, ganglion cell complex and the anterior chamber angle. The advantages and fallacies of OCT are discussed with several clinical examples for an easy learning experience. Common pitfalls in OCT imaging and interpretation are mentioned along with an extensive review of literature. I hope that the reader will benefit from this practical manual and gain expertise in the use of OCT for the diagnosis and progression of glaucomatous optic neuropathy.

Tanuj Dada

Acknowledgments

I would like to thank **Prof. RV Azad**, Chief, Dr Rajendra Prasad Centre for Ophthalmic Sciences, New Delhi, India, for his constant guidance and encouragement. He has been instrumental in creating a world class work environment for me.

I am very lucky to be in this great temple of learning with unmatched facilities for clinical care and research. I would also like to thank **Prof. Ramanjit Sihota** for her leadership in making the glaucoma services at Dr Rajendra Prasad Centre for Ophthalmic Sciences, New Delhi, India, one of the best in the world.

Tanuj Dada

Contents

1. **Introduction**   1

2. **Principle of Optical Coherence Tomography**   9

3. **Optical Coherence Tomography in Glaucoma**   13
    Optic Nerve Head Analysis   13
    Retinal Nerve Fiber Layer Analysis   14
    Guided Progression Analysis   16
    Ganglion Cell Analysis   17
    Macular Analysis   17
    Anterior Segment Evaluation   17
    Anterior Segment *Visante* OCT   17
    Lamina Cribrosa Imaging   20

4. **Optic Nerve Head Analysis**   24
    Technology   25
    Method   25
    Analysis   25

5. **Retinal Nerve Fiber Analysis**   33
    Method   34
    Technology   34
    Analysis   34
    Interpretation   35
    CASES
    Case 1: Structure Correlates with Function   43
    Case 2: Preperimetric and Perimetric Glaucoma   45
    Case 3: Structural Change Correlates with Function (Ignore Artifacts)   47
    Case 4: RNFL Thinning could be due to Nonglaucomatous Causes   49
    Case 5: Understanding the TSNIT RNFL Profile   50
    Case 6: Qualitative Analysis   52

6. **Progression Analysis**   55
    Technology   55
    Method   57
    Interpretation   58

CASES
Case 1: OCT versus Fundus Photography 61
Case 2: Pseudoprogression 68
Case 3: Variability on OCT 76
Case 4: True Progression: OCT Correlates with Visual Fields 80

7. Ganglion Cell Analysis 87

Technology 88
Method 88
Analysis 88

CASES
Case 1: Understanding the Ganglion Cell Analysis.
   Always Compare with RNFL Analysis 92
Case 2: Advanced Glaucoma 94
Case 3: Preperimetric and Perimetric Glaucoma 96

8. Macular OCT 100

Technology 100
Method 103
Analysis 103

9. Anterior Segment Analysis 106

Technology and Use 106

10. Limitations of OCT and Imaging Artifacts 113

Acquiring a Good Scan: Pearls 113
Pitfalls in OCT Imaging 115
High Myopia 118

Index 125

CHAPTER

1

Introduction

Glaucoma is a progressive optic neuropathy characterized by pathological loss of retinal ganglion cells and their axons, seen in the form of the retinal nerve fiber layer (RNFL) loss. Clinically glaucoma is diagnosed on the basis of loss of neuroretinal rim and increased cupping of the optic nerve head. This along with wedge shaped nerve fiber layer defects helps clinch the diagnosis of glaucoma. These changes may or may not be associated with visual field loss. Considering that the diagnosis of glaucoma is almost entirely a clinical one, one could wonder where the optical coherence tomography (OCT) fits in the whole picture. However our understanding that structural changes of the optic nerve and nerve fiber layer are the bottom line in glaucoma implies that the OCT which is able to measure such changes at micron levels should theoretically perform better than the clinician for diagnosis of glaucoma. However while this holds true for a majority of the early and moderate glaucomas, it is not always the case, particularly in the more advanced cases. Another aspect is that functional changes take time to appear and it is seen that a large part of the RNFL is lost before a change is seen on the visual fields.[1] The loss of Retinal ganglion cells is irreversible; therefore early detection of its presence and progression is critical to management of glaucoma. The best tool to enable this is therefore the OCT and this defines its major role in the disease.

The optic nerve like most central nervous system tissue has immense redundancy and usually more than 60% of the RNFL is needed to be lost before a defect would be visible on the visual fields. It has been previously proven that structural loss precedes functional loss.[1-3] The appropriate method to determine structural loss is visualizing the optic nerve head and retinal nerve fiber layer changes and documenting the change in them through serial fundus photographs or diagrams. However such a procedure carries with it an element of subjectivity and is more of a qualitative measure than an objective quantitative one.

With the advent of the OCT in 1990 and its subsequent use in biological tissue by Fujimoto in 1991 and subsequently in the field of ophthalmology by Schuman, there has been a tremendous revolution in ophthalmic care. The resolution of the machine has been enhanced from 30µ to submicron levels, as the reliability of the readings continues to improve in parallel. The newer machines are more patient friendly, can acquire the image much quicker and perform complicated analysis to provide whatever report the ophthalmologist may require.

Much like the already existing Heidelberg Retinal Tomogram, the OCT was first designed and intended for use primarily to evaluate the retina and manage retinal pathologies. However soon its use grew to evaluate the RNFL in cases of optic neuropathies such as glaucoma and this has been furthered in the recent past by a module for evaluation of the ganglion cell layer.

The OCT has gone beyond imaging the posterior segment and one of the latest entrants in the line-up is the new anterior segment OCT which enables excellent visualization of the angle and ciliary body.

There are various tools to image the optic nerve head and the retinal nerve fiber layer. Each of these is based on different technologies to enable imaging including laser interferometry (OCT), scanning laser ophthalmoscopy (HRT) and scanning laser polarimetry (GDx). A wide variety of studies has compared OCT to other imaging modalities and found a good sensitivity and specificity and while each of these imaging modalities such as the GDx and the Heidelberg retinal tomogram have their own advantages and disadvantages, the OCT continues to be a popular choice.[4,5]

The OCT has found to be more advantageous than GDx and HRT when it comes to evaluating localized RNFL defects and the local peripapillary area changes.[6,7]

Table 1.1 compares and contrasts the various imaging modalities for glaucoma and highlights the advantages of each.

Figures 1.1 to 1.4 show the various imaging modalities available for glaucoma diagnosis and include a sample printout of each technology for understanding the different parameters studied in each. This manual discusses the use of OCT in glaucoma diagnosis and management and provides a practical approach to reading and interpreting an OCT analysis with the aim of maximizing the information gained from this investigation. Various clinical examples have been included to enable a realistic learning. In general, the chapters discuss the significance of each form of imaging protocol, the technology behind it, the technique of performing the scan and interpretation of the final output. Clinical cases help consolidate the interpretation of results.

Table 1.1 Salient features of the three common methods of disc imaging namely, optical coherence tomography, GDx and Heidelberg retinal tomography with particular reference to glaucoma.[1,2,5,8,9]

Imaging modality	Principle	Information obtained	Advantages	Drawbacks	Clinical uses
OCT	Uses low coherence infrared light (830 nm). Principle of Michelson interferometry	Optical disc scan/Optic nerve head scan Retinal nerve fiber layer analysis Can also examine macula to look for hypotonous maculopathy	• Objective, quantitative, reproducible measurements of the retina and RNFL thickness • Measurements are not affected by refractive status, axial length of the eye, or the presence of early-to-moderate nuclear sclerotic cataracts • Structural information is independent of any arbitrarily defined reference plane	• High cost • Requirement of pupillary dilation • Posterior subcapsular and cortical cataracts impair performance • Images contain significantly fewer pixels than both GDx and HRT	• RNFL evaluation in preperimetric glaucoma • Studying macular changes in hypotony • Evaluating cystoid macular edema surgery and use of anti-glaucoma medications. • Evaluating optic nerve head tomography in glaucoma patients.
GDx-VCC	Near infrared light (780 nm) Scanning laser Polarimetry	RNFL: Thickness and deviation Nerve fiber indicator	• Does not require pupillary dilatation • Good reproducibility • Does not require a reference plan	• Measures inferred RNFL thickness • Measures RNFL at different locations in different eyes • Does not differentiate true biological change from normal variability • Limited Indian population data • Affected by anterior and posterior segment pathology	Diagnosis of preperimetric glaucoma and follow-up Nerve fiber indicator defines the overall neural intergrity
HRT II	Diode laser light 670 nm Confocal laser Scanning ophthalmoscope Stereometric photography and use of Moorfield regression analyser	Retinal nerve fiber layer thickness Height variation contour	• Pupillary dilatation not required • Rapid scanning time • Eliminates out-of-focus images degrading data • Good reproducibility • Accurate progression detection	• Requires reference plane for all measurements • Reference plane can be tricked by tilted discs, bean pot cupping, and drusen, optic nerve physiological cupping • Indirect measure of RNFL thickness • Normative database based on few caucasian patients	Useful for diagnosing progression in preperimetric glaucomas

A

OD	Scans used	1, 2, 3
OS	Scans used	1, 2, 3

	OD (N=3)	OS (N=3)	OD-OS
Imax/Smax	1.19	1.08	0.11
Smax/Imax	0.84	0.93	-0.09
Smax/Tavg	4.21	2.95	1.26
Imax/Tavg	4.99	3.18	1.81
Smax/Navg	2.28	2.54	-0.26
Max-Min	102.00	171.00	-69.00
Smax	105.00	206.00	-101.00
Imax	125.00	222.00	-97.00
Savg	73.00	172.00	-99.00
Iavg	87.00	174.00	-87.00
Avg.Thick	57.94	124.04	-66.10

B

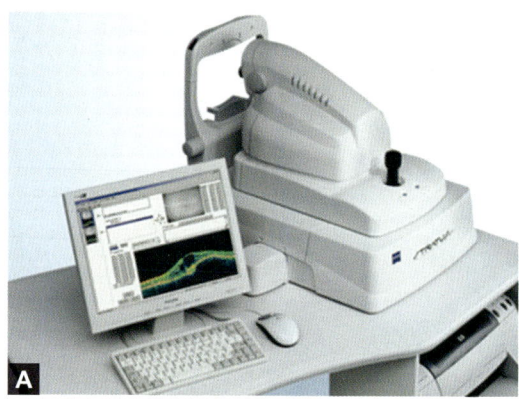

Figs 1.1A and B (A) Photograph of the Stratus Optical Coherence Tomography machine; (B) A sample printout of the RNFL analysis screen

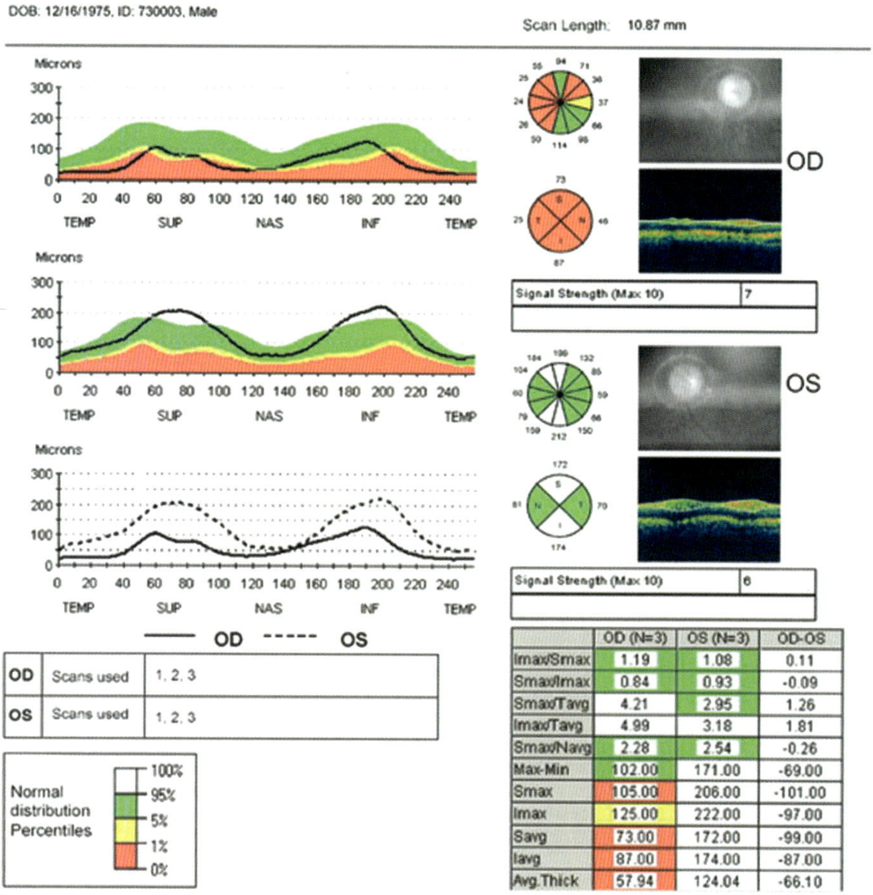

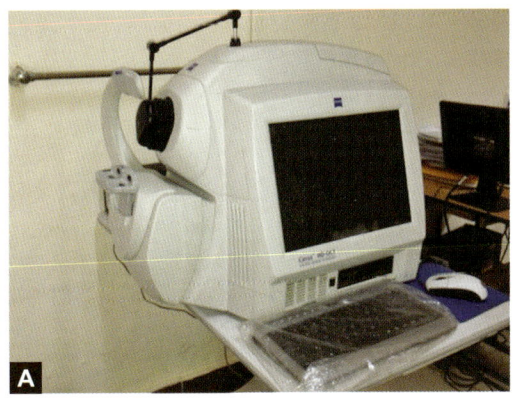

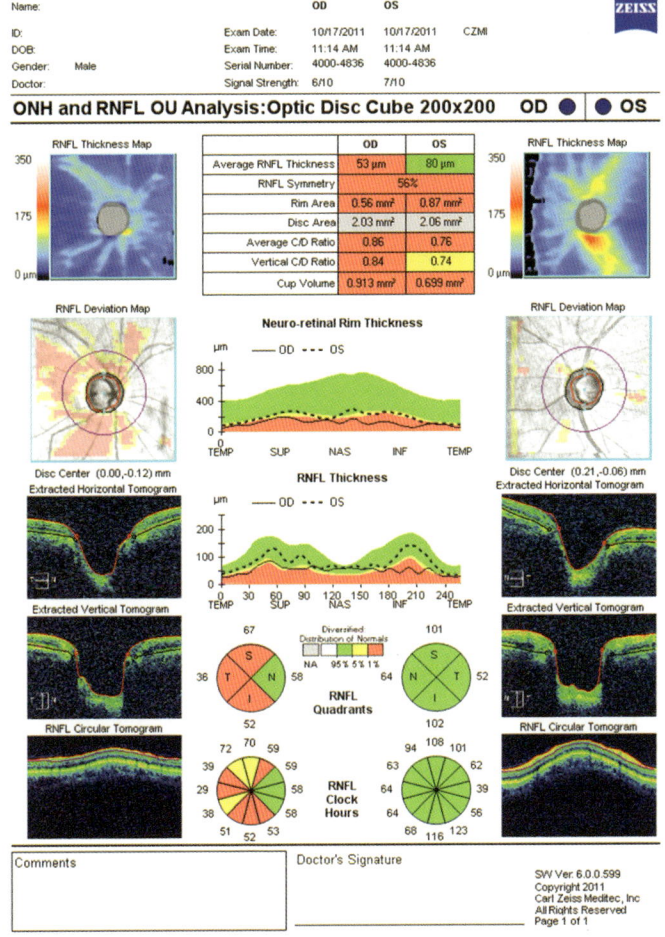

Figs 1.2A and B (A) Photograph of the Cirrus Optical Coherence Tomography machine; (B) A sample printout of the RNFL analysis screen

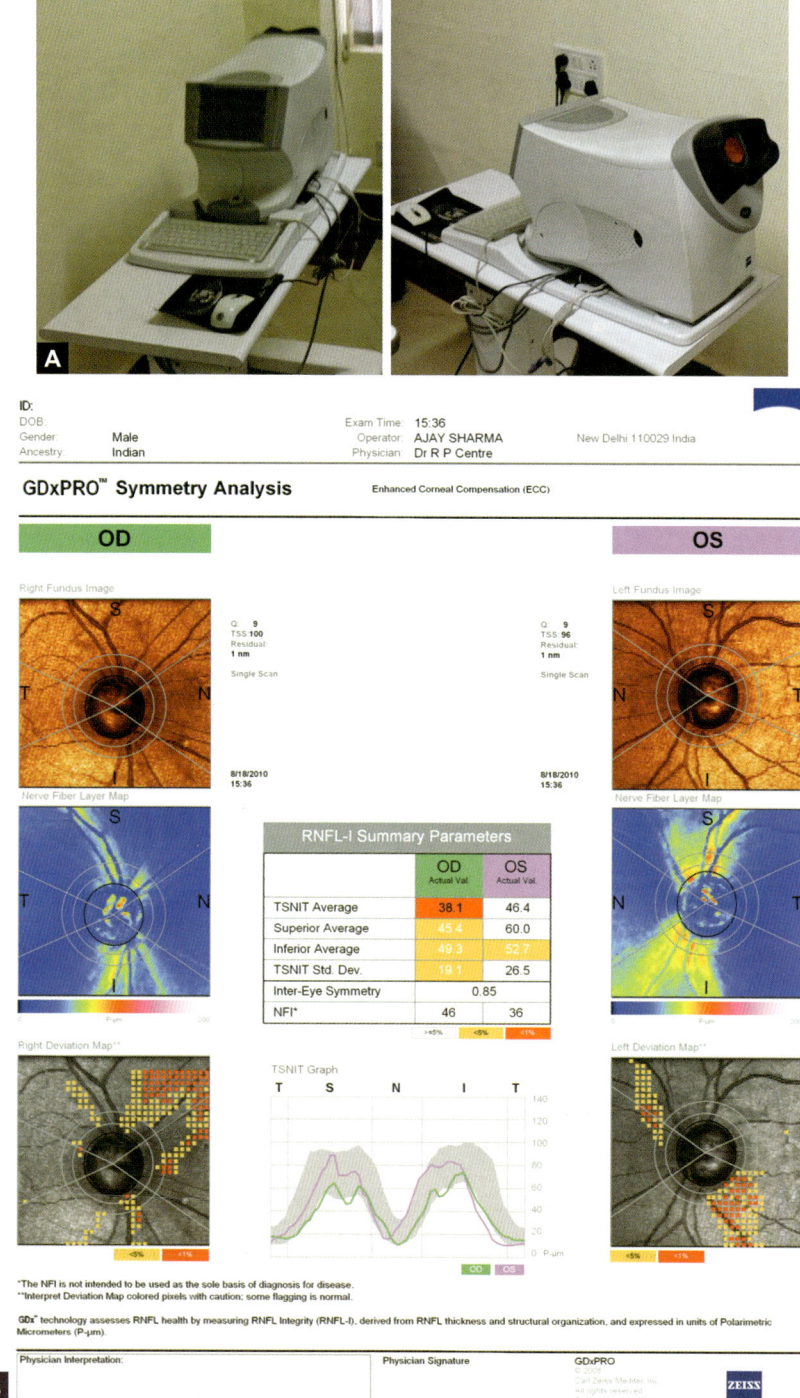

Figs 1.3A and B Photograph of the GDx-VCC machine and a sample printout of the RNFL analysis screen

A

Heidelberg Retina Tomograph
Initial Report

HEIDELBERG ENGINEERING

Patient:

Sex: male DOB: Pat-ID: Ethnicity: Indian

OS

Examination: **Date: Jan/23/2001**

Scan: Focus: 0.00 dpt Depth: 3.00 mm Operator: --- IOP: ---

Stereometric Analysis ONH		Normal Range
Disc Area	1.63 mm²	1.53 - 2.63
Cup Area	0.54 mm²	0.15 - 1.04
Rim Area	**1.09 mm²**	**1.22 - 1.75**
Cup Volume	0.08 mm³	-0.02 - 0.27
Rim Volume	**0.28 mm³**	**0.24 - 0.52**
Cup/Disc Area Ratio	0.33	0.13 - 0.39
Linear Cup/Disc Ratio	0.58	0.36 - 0.63
Mean Cup Depth	0.25 mm	0.13 - 0.28
Maximum Cup Depth	0.62 mm	0.40 - 0.73
Cup Shape Measure	**-0.14**	**-0.25 - -0.10**
Height Variation Contour	**0.43 mm**	**0.29 - 0.49**
Mean RNFL Thickness	**0.25 mm**	**0.19 - 0.32**
RNFL Cross Sectional Area	1.15 mm²	0.94 - 1.61
Reference Height	374 µm	
Topography Std Dev.	12 µm	
FSM	-0.50	
RB	0.95	

Moorfields Classification: Borderline (*)

Comments:

Date: Aug/30/2013 Signature:

B

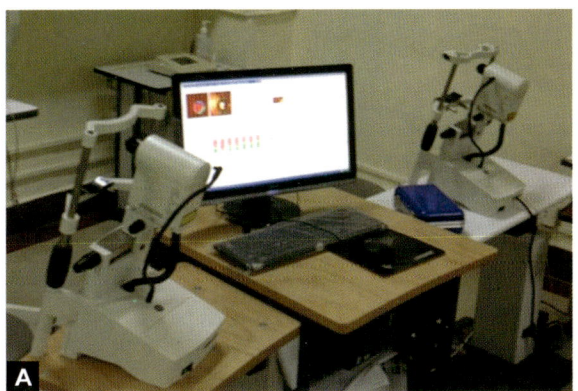

Figs 1.4A and B Photograph of the Heidelberg Retinal Tomography machine and a sample printout of the RNFL analysis screen

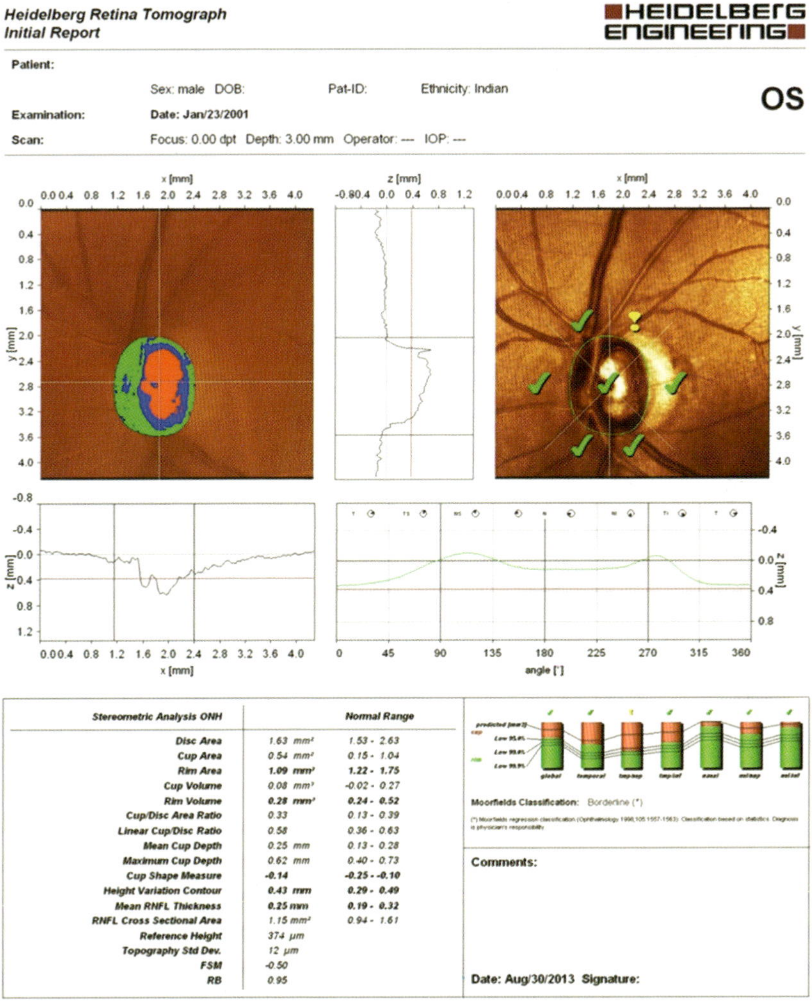

Key Points

- Structural changes in optic nerve head and RNFL precede functional changes on automated perimetry.
- OCT can detect RNFL changes prior to visual field loss.
- OCT, SLP and CSLO are the current imaging modalities for evaluating glaucoma related changes in the optic nerve head and RNFL.
- OCT is the only imaging modality which provides high resolution images of optic nerve head, RNFL, macula, ganglion cell complex and anterior chamber angle.

■ REFERENCES

1. Gordon MO, Beiser JA, Brandt JC, et al. The ocular hypertension treatment study: Baseline factors that predict the onset of primary open-angle glaucoma. Arch Ophthalmol 2002;120:714-12.
2. Medeiros FA, Zangwill LM, Bowd C, Mansouri K, Weinreb RN. The structure and function relationship in glaucoma: implications for detection of progression and measurement of rates of change. Invest Ophthalmol Vis Sci. 2012;53(11):6939-46.
3. Graham S. Defining the structure/function relationship in glaucoma. Clin Experiment Ophthalmol. 2012;40(4):337-8.
4. Sehi M, Bhardwaj N, Chung YS, Greenfield DS. Advanced Imaging for Glaucoma Study Group. Evaluation of baseline structural factors for predicting glaucomatous visual field progression using optical coherence tomography, scanning laser polarimetry and confocal scanning laser ophthalmoscopy. Eye (Land). 2012;26(12):1527-35.
5. Sanchez-Garcia M, Rodriguez de la Vega R, Gonzalez-Hernandez M, Gonzalez de la Rosa M. Variability and reproducibility of 3 methods for measuring the thickness of the nerve fiber layer. Arch Soc Esp Oftalmol. 2013;88(10):393-7.
6. Na JH, Lee KS, Lee JR, Lee Y, Kook MS. The glaucoma detection capability of spectral domain OCT and GDx-VCC deviation maps in early glaucoma patients with localized visual field defects. Graefes Arch Clin Exp Ophthalmol. 2013;251(10):2371-82.
7. Wang H, Tao Y, Sun XL, Zhuang K. Comparison of Heidelberg retina tomography, optical coherence tomography and Humphrey visual field in early glaucoma diagnosis. J Int Med Res. 2013;41(5):1594-605.
8. Pueyo V, Polo V, Larrosa JM, Ferreras A, Pablo LE, Honrubia FM. Diagnostic ability of the Heidelberg retina tomograph, optical coherence tomograph, and scanning laser polarimeter in open-angle glaucoma. J Glaucoma. 2007;16(2):173-7.
9. Pueyo V, Polo V, Larrosa JM, Ferreras A, Alias E, Honrubia FM. Ability of different optical imaging devices to discriminate between healthy and glaucomatous eyes. Ann Ophthalmol (Skokie). 2009;41(2):102-8.

CHAPTER 2

Principle of Optical Coherence Tomography

Optical coherence tomography (OCT) is based on the principle of Michelson interferometry.[1] Low-coherence infrared (830 nm) light is projected on a beam splitter and one of the beams is thereon directed through the ocular media to the retina while the other beam is focused onto a reference mirror. Light passing through the eye is reflected by the retinal structures (Figure 2.1). Meanwhile, the distance between the beam-splitter and reference mirror is continuously varied. When the distance between the light source and retinal tissue is equal

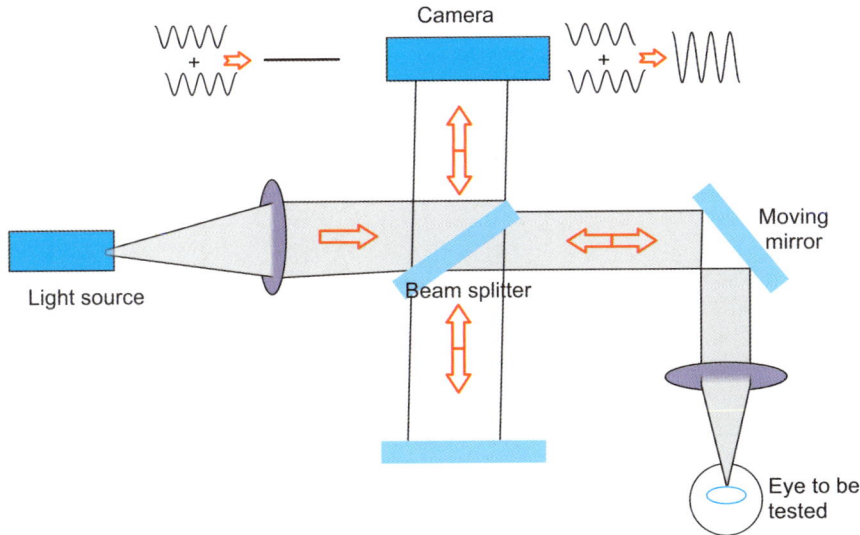

Fig. 2.1 The principle of the OCT machine

to the distance between the light source and reference mirror, the reflected light from the retinal tissue and reference mirror interacts to produce an interference pattern. The interference pattern is detected by a camera detector array and then processed into a signal. The processed signal, reflected from various points on the retina, forms a two-dimensional image of the retina. The image is, therefore, a series of aligned A-scans that resemble a histologic section. The beam reflected from the retina is captured by an infrared sensitive charge-coupled device video camera along with an infrared fundus photograph defining the area of retina captured. The newer spectral or fourier domain OCT uses a broad spectrum of light source to produce multiple interference patterns at different depths and eliminate the moving mirror to provide a higher resolution and faster scan.

The light source employed in the OCT is a low coherence super luminescent diode source (820 nm) for the time domain *Stratus* OCT and a titanium sapphire laser (broad bandwidth centered at 800 nm) for the fourier domain *Cirrus* OCT. Spectral-domain technology differs significantly from time-domain technology in that signals returning from the eye are scanned by a spectrometer to analyse alteration of light frequency compared with input frequency. Faster scanning and data collection are possible which result in many improvements, such as less patient-movement artifact and markedly improved image registration.[2] With more data acquired in each scan session, less interpolation between scans is required, making volumetric analysis and 3 D imaging possible. Measurements are reproducible within reasonable tolerances, and registration aids in establishing proper sites for measurement.

The spectral-domain OCT device can produce cross-sectional B-scans, like time-domain OCT but with better resolution, and it can also create 3D area scans by combining B-scans. Its scanning technology takes 20,000 to 26,000 A-scan measurements per second, produces a linear B-scan in less than 0.03 second, and combines them to create a 3 D area scan.[2] The differences between the time domain and spectral domain OCT are depicted in Table 2.1.

The OCT image can be displayed either on a gray scale where more highly reflected light is brighter than less highly reflected light or in color where each color represents a degree of reflectivity (highly reflective structures are shown with bright colors (red and yellow), while those with low reflectivity are represented by darker colors (black and blue). Those with intermediate reflectivity appear green. Optical coherence tomography can be used to visualize changes in tissue optical scattering properties or refractive index discontinuities, but it cannot distinguish between tissues of similar optical properties. Clinical practice of glaucoma is advancing in every aspect. One of these includes the use of the spectral domain OCT instead of the time domain OCT. Literature has shown that the spectral domain OCT has better sensitivity and specificity (63.6% and 100% respectively) than time domain OCT (40.0% and 96.7% respectively) with fewer artefacts (35 vs 26%).[3,4] However, certain studies in literature have shown that the results achieved by both OCT's are similar and correlate well with each other.[5,6] However, it is recommended to follow up the patients on any one device for better reliability and reproducibility.

Table 2.1 Comparing the time domain and spectral domain OCT

Time domain OCT	Fourier/Spectral domain OCT
Stratus OCT	Cirrus OCT or HROCT
Mechanical moving part (Mirror)	*No moving part:* Broadband source is employed
Information along the longitudinal direction is accumulated through the course of the longitudinal scan time	The entire signal and information is recorded in parallel by a spectrometer
Light source: Super luminescent diode source (wavelength 810 nm, bandwidth 20 nm)	*Light source:* Super luminescent diode produces a light with a wavelength of 840 nm (bandwidth 150 nm)
Longer scan acquisition time	50–100 times faster (shorter scan times)
Eye movement affects scan quality significantly	Eye movement does not affect scan quality as much
400 axial scans per second	25,000–100,000 axial scans per second
1 B scan with 512 A-scans taken per B scan	256 B scans taken
8-10 µm axial resolution	3–7 µm axial resolution
End point of measuring retinal thickness measured between inner limiting membrane and the inner-outer segment junction of photoreceptor	End point of measuring retinal thickness between inner limiting membrane and the outer boundary is the retinal pigment epithelium (outer boundary not clearly known)
Three-dimensional reconstruction not possible	Three-dimensional reconstruction is possible as the consecutive A scans do not require digital alignment

Key Points

- OCT is based on the principle of Michelson interferometry
- Spectral domain OCT has a better resolution than time domain OCT.
- Spectral domain OCT has a shorter acquisition time as compared to time domain OCT and can provide 3-D rendering of the images.
- End point of measuring retinal thickness in spectral domain OCT and time domain OCT are different.

REFERENCES

1. Jaffe GJ, Caprioli J. Optical coherence tomography to detect and manage retinal disease and glaucoma. Am J Ophthalmol. 2004;137(1):156-69.
2. Wojtkowski M, Bajraszewski T, Gorczyńska I, Targowski P, Kowalczyk A, Wasilewski W, Radzewicz C. Ophthalmic imaging by spectral optical coherence tomography. Am J Ophthalmol. 2004;138(3):412-9.

3. Sung KR, Kim DY, Park SB, Kook MS. Comparison of retinal nerve fiber layer thickness measured by Cirrus HD and Stratus optical coherence tomography. Ophthalmology. 2009;116(7):1264-70, 1270.e1.
4. Forte R, Cennamo GL, Finelli ML, Crecchio G de. Comparison of Time Domain Stratus OCT and Spectral Domain SLO/OCT for Assessment of Macular Thickness and Volume. Eye. 2009;23(11):2071-8.
5. Park KH, Jeoung JW. Comparison of Cirrus OCT and Stratus OCT on the Ability to Detect Localized Retinal Nerve Fiber Layer Defects in Preperimetric Glaucoma. Invest Ophthalmol Vis Sci. 2010;51(2):938-45.
6. Moreno-Montañés J, Olmo N, Alvarez A, García N, Zarranz-Ventura J. Cirrus high-definition optical coherence tomography compared with stratus optical coherence tomography in glaucoma diagnosis. Invest Ophthalmol Vis Sci. 2010;51(1):335-43.

CHAPTER

3

Optical Coherence Tomography in Glaucoma

Optical coherence tomography has been used in various aspects of diagnosis and follow-up in glaucoma patients. The various facets of glaucoma evaluated by OCT include:[1,2]
- Optic nerve head analysis
- Retinal nerve fiber layer thickness measurement
- Guided progression analysis
- Ganglion cell layer thickness
- *Macular analysis:* Thickness and profile
- Anterior Segment examination
- *Anterior Segment-OCT:* Visante OCT
- Lamina cribrosa.

This book predominantly focuses on the posterior segment OCT and the modules available for commercial use. The chapters in this book discuss in detail each of the important aspects that may be studied using the posterior segment OCT (*Stratus* and *Cirrus* OCT).

■ OPTIC NERVE HEAD ANALYSIS

Changes in the optic nerve head is the mainstay for the clinical diagnosis of glaucoma. The prominent components of the optic nerve head analysis (Figure 3.1) which are seen clinically include the neuro-retinal rim profile and thickness and the amount of cupping. Even on the OCT, the primary parameters seen are the neuro-retinal rim area, cup area, integrated rim area and the cup volume among other parameters. It is important to understand that these parameters are entirely objective and are average values, not defined by quadrant or sector unlike a clinical examination. Qualitative assessment of rim notches or depth is not accurately possible. The optic nerve head along with advances in the form of three dimensional reconstruction and progression analysis forms an integral part of the role of OCT in glaucoma. More details on optic nerve head analysis are in chapter 4.

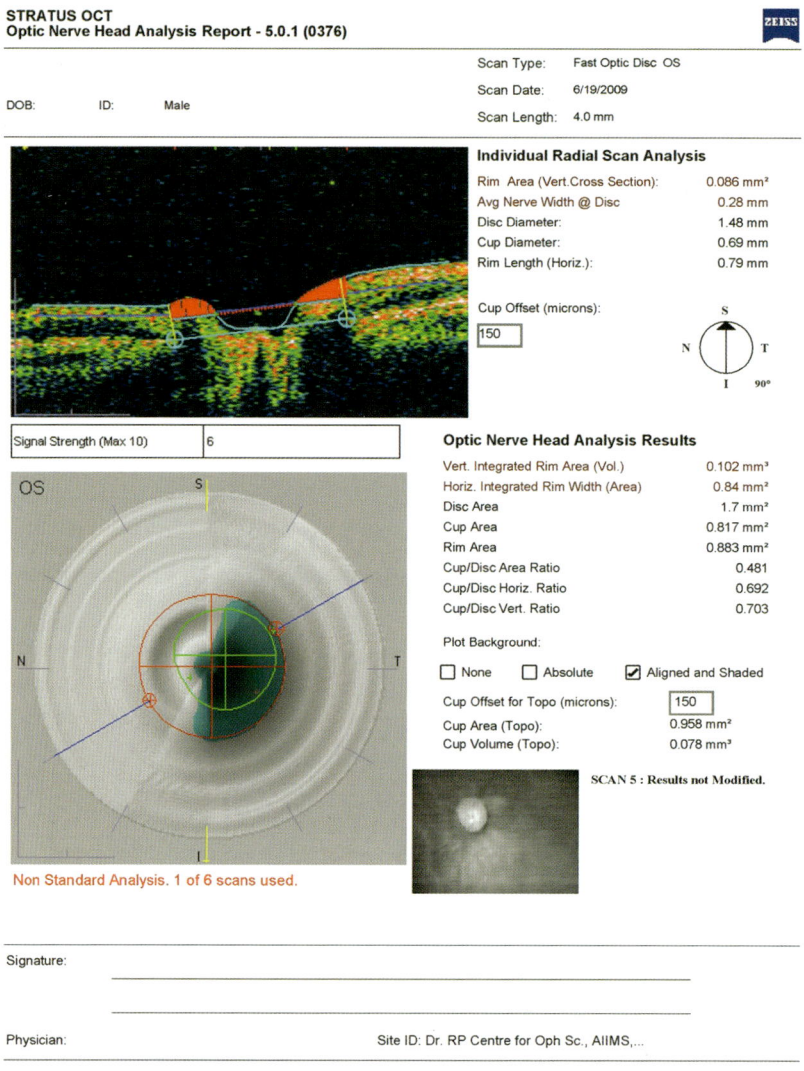

Fig. 3.1 Typical printout of an optic nerve head analysis

RETINAL NERVE FIBER LAYER ANALYSIS

Probably the most important role of the OCT in diagnosing and following up glaucoma has been its ability to qualify RNFL and progressively follow its changes. It has been seen that RNFL change has preceded field change and this enables early diagnosis. This parameter is accuartely measured and has been shown to be highly reproducible and reliable for follow-up. The importance of sector-wise and quadrant-wise analysis was understood when studies demonstrated certain sectors or clock hours to get affected preferentially in glaucoma, particularly in progressive disease. The important aspects of RNFL studied include RNFL thickness, its deviation from the normative database and its profile over the temporal, superior, nasal and inferior quadrants (TSNIT profile) (Figure 3.2). Infact, the OCT RNFL analysis has become such an important tool to enhance our diagnostic ability of glaucoma that over time, it given rise to the term red disease referring to glaucoma diagnosed on optical coherence tomography (OCT) but not on clinical or functional tests. A wide variety of studies has compared optical coherence tomography (OCT) to other imaging modalities.

Further details on RNFL analysis are given in chapter 5.

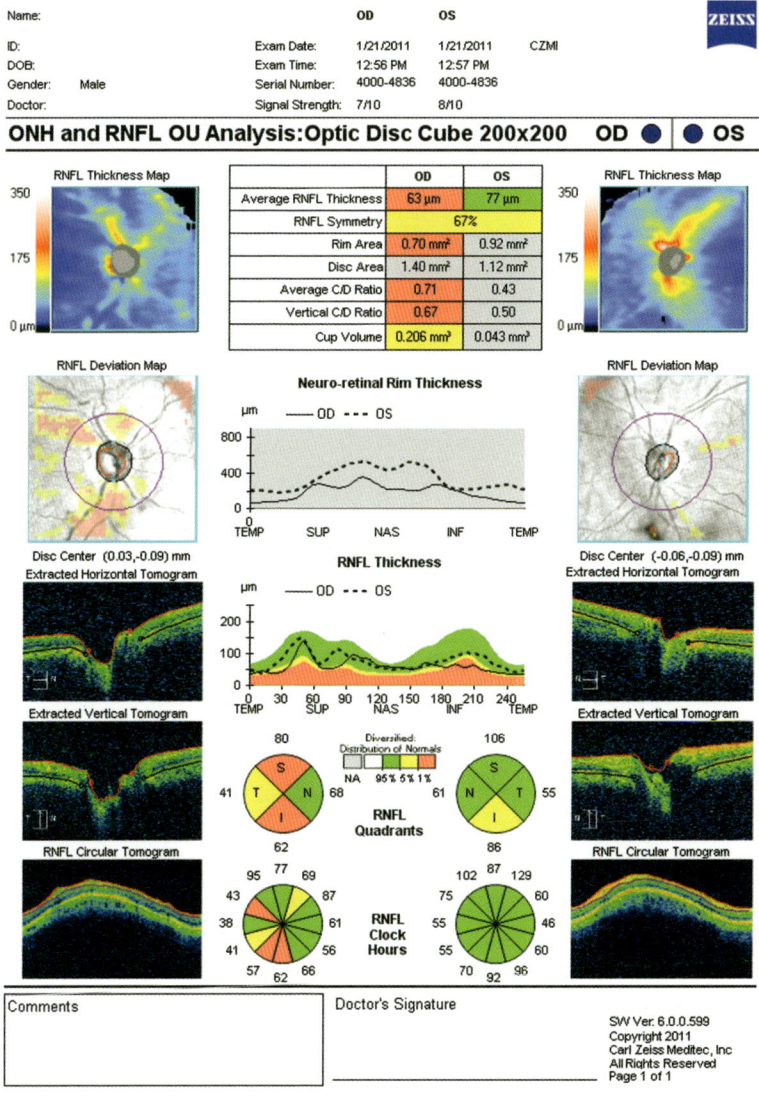

Fig. 3.2 Typical printout of a RNFL analysis

GUIDED PROGRESSION ANALYSIS

One of the main advantages the OCT has provided over progressive fundus images is the ability to measure and quantify change. This quantification has been useful as it has enabled a statistical test to determine whether progression has occurred and its significance. There are many cases, particularly in early glaucoma where progression is diagnosed only on structural change as these are preperimetric. Further, the ability to differentiate pathologic change from the normal physiologic change is not possible on serial fundus examinations. The GPA printout can be intimidating to read and understand and is often misinterpreted (Figure 3.3).

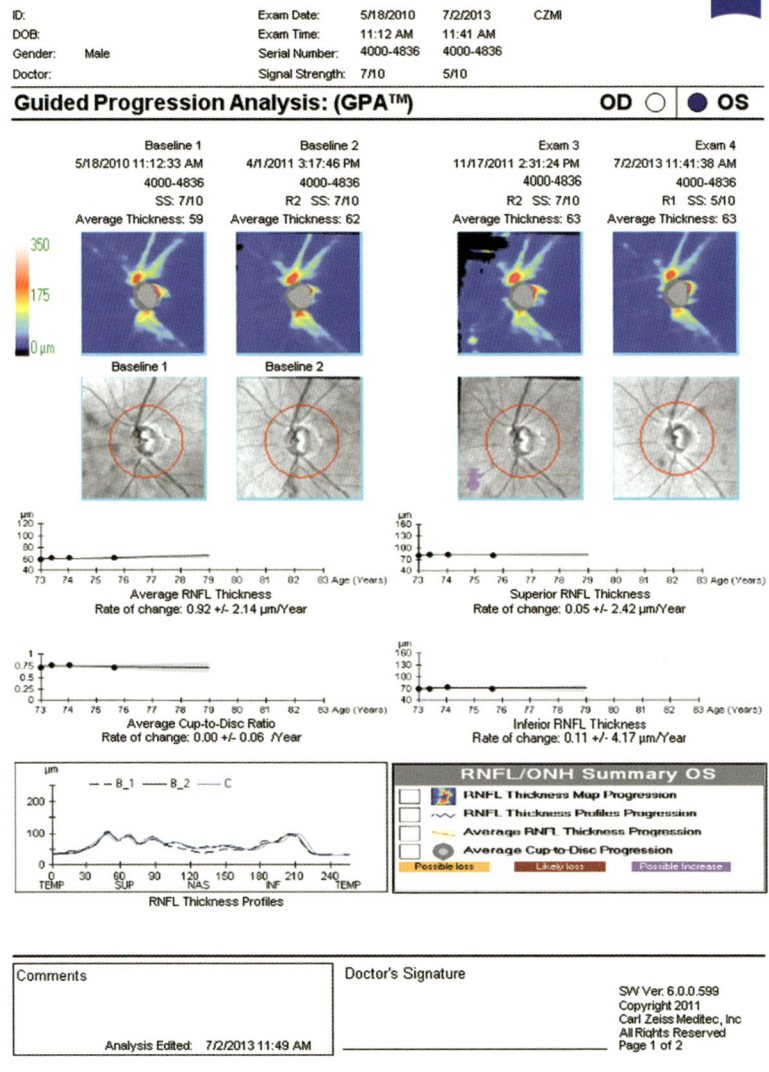

Fig. 3.3 Typical printout of guided progression analysis

Therefore chapter 6 discusses the guided progression analysis module of the OCT and provides a step by step guide to interpreting a GPA printout.

■ GANGLION CELL ANALYSIS

The ganglion cell analysis is a relatively recent concept that helps in measuring in an indirect manner the quantity of ganglion cells in the retina. This concept behind this module is that optic neuropathies, including glaucoma are primarily affecting the axons of the ganglion cell, therefore a measure of these would be a direct indication of glaucomatous damage. This parameter has shown a sensitivity similar to RNFL changes for most cases, but has better sensitivity for structural changes resulting in paracentral visual field loss. The ganglion cell analysis is acquired from the macular cube scan through a special software module and currently lots of research is ongoing on this aspect (Figure 3.4). Ganglion cell analysis is discussed in detail in chapter 7.

■ MACULAR ANALYSIS

It might seem that the macula does not have much role in glaucoma evaluation and its management but studies have shown otherwise. The important parameters evaluated is the macular profile where hypotonic maculopathy or cystoid macular edema (post-glaucoma surgery or secondary to prostaglandins) may be diagnosed and macular thickness which has been shown to correlate with presence of glaucoma. Macular scanning may be done using either a macular cube or a 5-line raster method and both provide important information relevant to glaucoma (Figure 3.5). The macular OCT is discussed in detail in chapter 8.

■ ANTERIOR SEGMENT EVALUATION

Glaucoma assessment involves two components. The first are the changes in the posterior segment including optic nerve head changes, nerve fiber changes and macular evaluation. The second is the assessment of the anterior segemnt to a evaluate the angle and cornea. The latter can also be performed using the OCT whereby a non-invasive imaging of the angle and the cornea or sclera can be done (Figure 3.6). This is possible due to a special anterior segment module in the machine software. The images are excellent and are free from any manipulation as may happen in gonioscopy or ultrasound biomicroscopy. While this is not the mainstay for the OCT in glaucoma assessment, it is an added feature that can often be helpful for an accurate diagnosis. This module is discussed in greater detail in chapter 9.

■ ANTERIOR SEGMENT *VISANTE* OCT

The anterior segment OCT (ASOCT) is one of the latest arsenal in glaucoma imaging and has enabled high resolution images of the cornea, angle, part of the iris and ciliary body with the aid of a super luminescence diode having an infrared wavelength of 1310 nm.[3,4] The major advantages of the ASOCT over

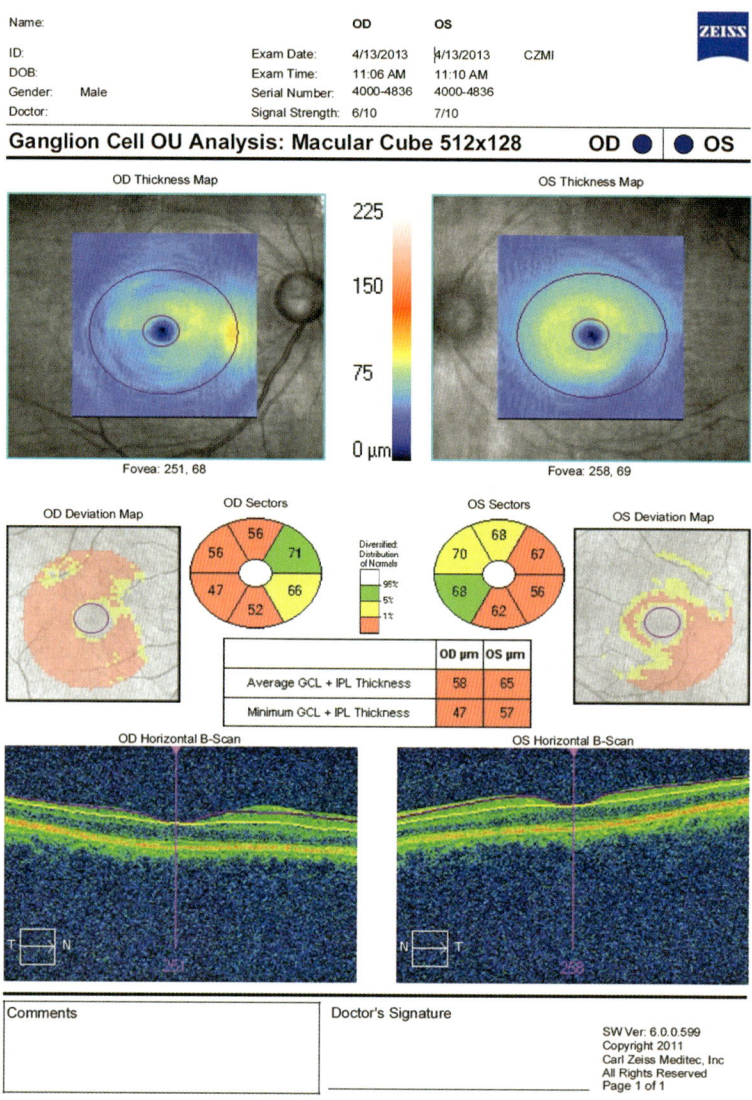

Fig. 3.4 Typical printout of a ganglion cell analysis

other forms of anterior segment imaging includes the ability to image a nascent angle unaffected by light (since it employs non-visible EM waves), the ability to image the anterior segment structures in sitting position, the ability of being more reproducible and above all being a non-invasive (hence enabling imaging in a postoperative patient). The drawbacks of the machine include the need for a reasonable clear medium for the IR rays to pass through for imaging and the need for the eye to be held still. Patients who can't be made to sit for an adequate period are also difficult to image.

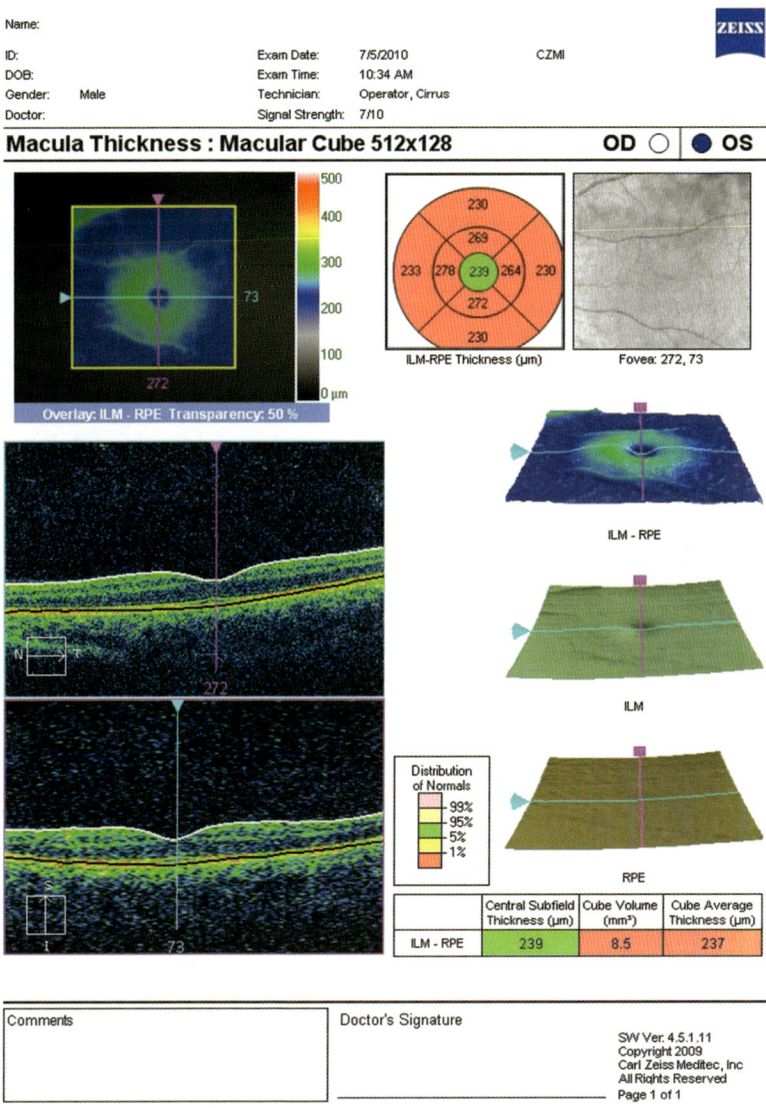

Fig. 3.5 Typical printout of a macular analysis

In glaucoma, the ASOCT can be used to determine angle closure including the degree of synechial closure versus appositional closure; it can determine the patency of the peripheral iridotomy and the affect it had on the angle; it can help diagnose tricky situations such as plateau iris or aqueous misdirection; it can help plan surgery in the preoperative period and in the postoperative period it can determine causes of failure or evidence of surgical success.[5] Detailed discussion and exemplification are beyond the scope of this book. Typical printout of a Visante OCT is shown in Figure 3.7.

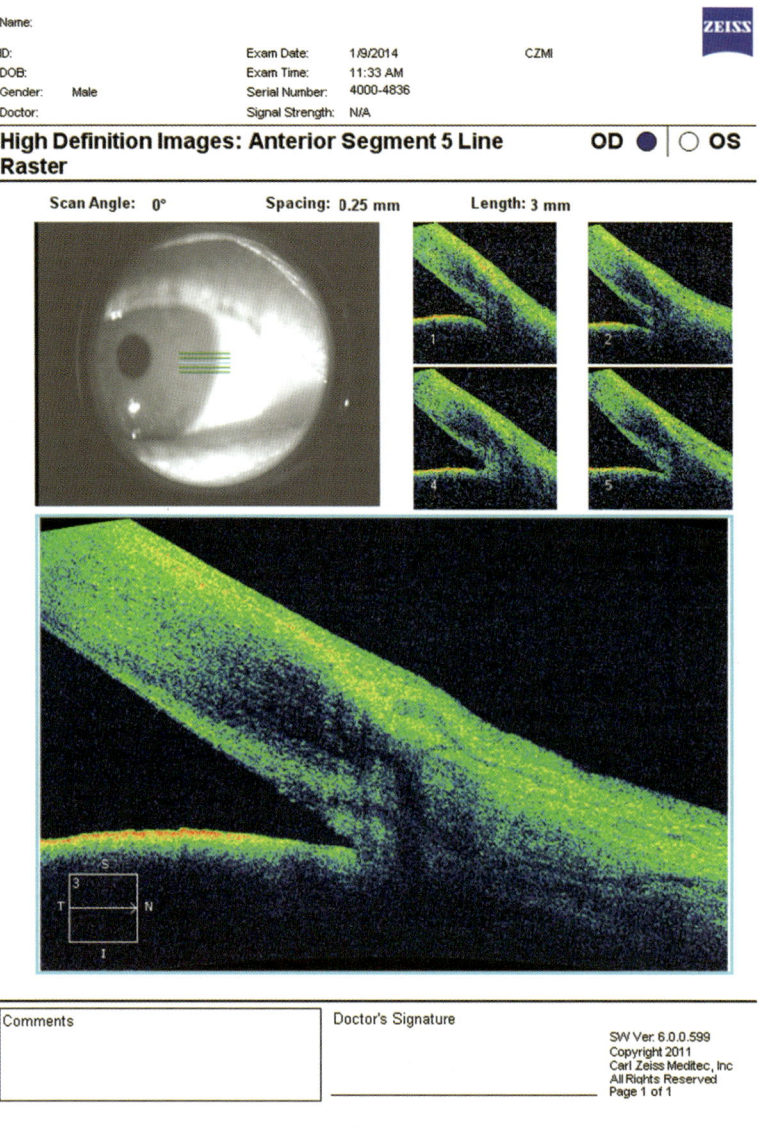

Fig. 3.6 Typical printout of an anterior segment analysis

■ LAMINA CRIBROSA IMAGING

The swept source OCT or OCT with in depth imaging or HR OCT have been used in the field of glaucoma research to view the lamina cribrosa for changes which may predispose an individual to glaucoma.[6,7] Viewing of the lamina cribrosa may aid in the early diagnosis of glaucoma or progression and this structure of the eye is now gaining importance as a site of pathophysiological changes in glaucoma.

While not yet in commercial use, recent literature has shown micro-architecture changes in which reflect beams' remodeling and axonal loss leading to reduction in pore size and increased pore size variability of the lamina cribrosa of glaucoma patients as compared to healthy controls and this showed a correlation with worse visual fields.[8] Postoperatively after glaucoma surgery, lamina cribrosa depth has shown to decrease and the prelaminar neural tissue has shown to get thickened in correlation to the decrease in intraocular pressure.[9] Other studies have found disc hemorrhage, a diagnosis of normal tension

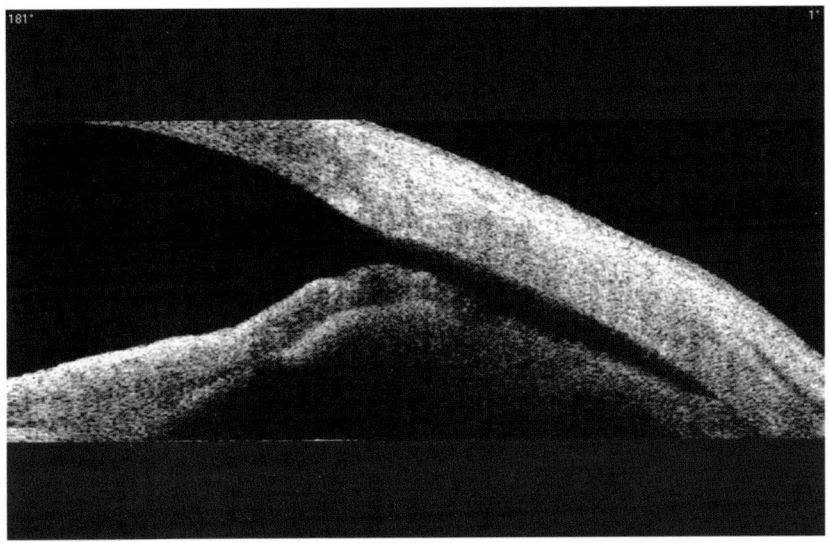

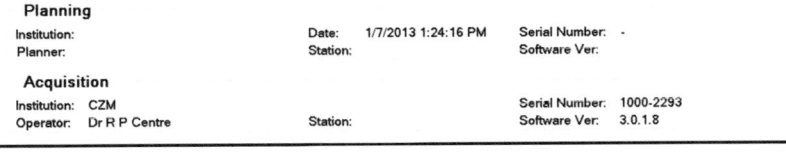

Fig. 3.7A

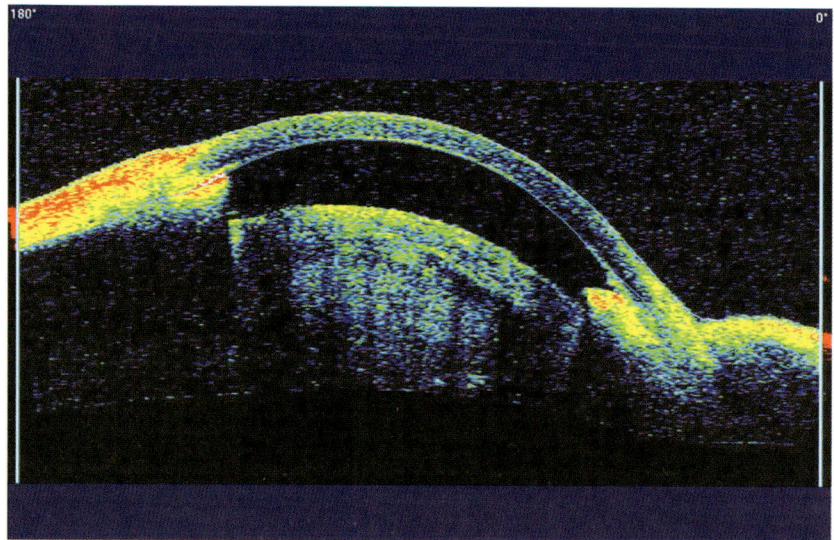

Fig. 3.7B

Figs 3.7A and B Typical printout of a Visante OCT anterior segment analysis. (A) Shows high resolution imaging of the anterior chamber angle using the Visante OCT. Note the high resolution and details; (B) Shows a lower resolution imaging of the cornea, anterior chamber and anterior surface of lens

glaucoma, RNFL defects and more advanced glaucoma status to be associated with focal lamina cribrosa defects.[10,11] A detailed discussion on this new facet of imaging is beyond the scope of this book.

The OCT has been an important tool to enhance our diagnostic ability of glaucoma and over the time, OCT and other imaging modalities has given rise to the term red disease referring to glaucoma diagnosed on OCT but not on clinical or functional tests.[12]

> **Key Points**
> - Structural progression is the gold standard for glaucoma diagnosis and progression
> - OCT is best method for RNFL analysis
> - Optic nerve head analysis, RNFL analysis and Ganglion cell analysis are the main stay for use of OCT in glaucoma
> - Macular scan and anterior segment scans provide additional information in glaucoma cases.

REFERENCES

1. Bussel II, Wollstein G, Schuman JS. OCT for glaucoma diagnosis, screening and detection of glaucoma progression. Br J Ophthalmol. 2013.
2. Geimer SA. Glaucoma diagnostics. Acta Ophthalmol. 2013.
3. Sharma R, Sharma A, Arora T, Sharma S, Sobti A, Jha B, Chaturvedi N, Dada T. Application of anterior segment optical coherence tomography in glaucoma. Surv Ophthalmol. 2013.
4. Pekmezci M, Porco TC, Lin SC. Anterior segment optical coherence tomography as a screening tool for the assessment of the anterior segment angle. Ophthalmic Surg Lasers Imaging. 2009;40(4):389-98.
5. How AC, Baskaran M, Kumar RS, He M, Foster PJ, Lavanya R, Wong HT, Chew PT, Friedman DS, Aung T. Changes in anterior segment morphology after laser peripheral iridotomy: an anterior segment optical coherence tomography study. Ophthalmology. 2012;119(7):1383-7.
6. Miki A, Ikuno Y, Jo Y, Nishida K. Comparison of enhanced depth imaging and high-penetration optical coherence tomography for imaging deep optic nerve head and parapapillary structures. Clin Ophthalmol. 2013;7:1995-2001.
7. Lopilly Park HY, Shin HY, Park CK. Imaging the Posterior Segment of the Eye using Swept-Source Optical Coherence Tomography in Myopic Glaucoma Eyes: Comparison With Enhanced-Depth Imaging. Am J Ophthalmol. 2013.
8. Wang B, Nevins JE, Nadler Z, Wollstein G, Ishikawa H, Bilonick RA, Kagemann L, Sigal IA, Grulkowski I, Liu JJ, Kraus M, Lu CD, Hornegger J, Fujimoto JG, Schuman JS. *In vivo* lamina cribrosa micro-architecture in healthy and glaucomatous eyes as assessed by optical coherence tomography. Invest Ophthalmol Vis Sci. 2013;54(13):8270-4.
9. Yoshikawa M, Akagi T, Hangai M, Ikeda HO, Takayama K, Morooka S, Kimura Y, Nakano N, Yoshimura N. Alterations in the Neural and Connective Tissue Components of Glaucomatous Cupping after Glaucoma Surgery using Swept Source Optical Coherence Tomography. Invest Ophthalmol Vis Sci. 2014.
10. Park SC, Hsu AT, Su D, Simonson JL, Al-Jumayli M, Liu Y, Liebmann JM, Ritch R. Factors associated with focal lamina cribrosa defects in glaucoma. Invest Ophthalmol Vis Sci. 2013;54(13):8401-7.
11. Tatham AJ, Miki A, Weinreb RN, Zangwill LM, Medeiros FA. Defects of the lamina cribrosa in eyes with localized retinal nerve fiber layer loss. Ophthalmology. 2014;121(1):110-8.
12. Kiddee W, Tantisarasart T, Wangsupadilok B. Performance of optical coherence tomography for distinguishing between normal eyes, glaucoma suspect and glaucomatous eyes. J Med Assoc Thai. 2013;96(6):689-95.

CHAPTER

4

Optic Nerve Head Analysis

■ INTRODUCTION

The OCT has become an important tool for optic nerve head analysis. With the increasing resolution and better reproducibility, the investigation is now almost considered a routine for evaluation of the optic nerve head especially in cases of glaucoma and optic neuropathies.[1-3] Specific to the evaluation of the optic nerve head, are cases of disc anomalies of neuro-ophthalmologic importance, cases with large discs and cases where the anatomy of a particular cupping needs to be examined (shallow versus deep or gentle versus steep slope of cupping, etc.) since these are not evident in an HRT or the GDx.[4] The OCT being an exact machine determined image and not one determined by the operator's decision regarding the origin of the cup, slope and neuroretinal rim, there is an advantage where different operators or examiners are involved.

The optic nerve head analysis has been shown to assist in accurate diagnosis of preperimetric glaucoma and progression of glaucoma, though the role of RNFL measurements is more in the latter case.[5,6] The latest advances in optic nerve head imaging have included enhanced and indepth imaging where lamina cribrosa and the posterior sclera have been studied (see chapter 3) to look for changes suggestive of glaucoma. Among other advances in optic nerve head imaging is the use of the 5-line raster scan to manually measure parameters like neural canal opening, prelaminar canal depth, peripapillary choroidal thickness, and canal nerve fiber layer which provide an insight into the detailed anatomy of the optic nerve head and possibly will aid in diagnosis of glaucoma.[7] Also, nerve head imaging to establish the true dimensions of the disc by determining the Bruch membrane opening and then deducing the minimum rim width and minimum rim area is postulated to improve diagnostic ability.[8] Various optic nerve head parameters have shown good sensitivity and specificity for glaucoma diagnosis. These include vertical rim thickness, rim area and vertical cup disc ratio having a sensitivity of 83.9%, 77.4% and 77.4% with specificities of 91.8%, 97.3% and 93.8%, respectively.[9]

TECHNOLOGY

The science and technology behind optic disc analysis is as follows. The termination of the Bruch's membrane defines the disc. The rim width around the circumference of the optic disc is determined by measuring the amount of neuro-retinal tissue in the optic nerve. This is a more true measure of the neuro-retinal rim unlike other techniques of disc imaging that determine the cup margin based on its intersection with a plane at a fixed distance above the disc. In the OCT method, the disc and rim area measurements are done in the same plane as the optic disc, while the 2D drawing on other methods is in the plane of the end-on image. Measuring the area in the same plane as the optic disc provides a result that better correlates with the anatomy as against measuring in the end-on view.

METHOD

After entering the complete identity and demographic data of the patient, the optic nerve head scan is chosen from the scan options. The patient is placed in a comfortable sitting position with his/her chin on the chinrest and the forehead touching the forehead stand. Once the optic nerve head scan is chosen, the instrument moves the fixation target to the temporal side of the fixation field for that eye (i.e. if the right eye is being tested, the fixation star will move to the right side of the field visible to the patient while viewing into the machine) and the disc appears in the center of the acquisition field. From the operators standpoint, the acquisition scan area appears as a square with perpendicularly crossing target lines and a central target circle which should be made to encircle the disc circumferentially. After centering the pupil, the scan is moved to get the disc within the projected square. This scan is then optimized and captured. The result output may be displayed either quantitatively (Figure 4.1) or qualitatively (Figure 4.2).

ANALYSIS

The optic nerve head scan can be analyzed objectively or a subjective visualization can be done. The SD OCT provides an opportunity to view the disc in 3D and this helps identify any pathologies of shape or form (Figures 4.3A and B). An estimate of the cupping can also be made from this scan though it would usually tend to overestimate the cup area as compared to that viewed by the clinician or captured on a fundus photograph.

In the objective scan analysis, various parameters are examined and displayed. (Figure 4.4). This analysis detects the anterior surface of the retinal nerve fiber layer (RNFL) and the retinal pigment epithelium (RPE). The cup perimeter is determined by automatic detection of certain reference points most of which are based on anatomical landmarks, on each side of the disc where the RPE ends. The software locates and measures the disc diameter by tracing a straight line between the disc reference points. The cup diameter is measured on a line parallel to the disc line which is offset anteriorly by 150 microns. On the basis of

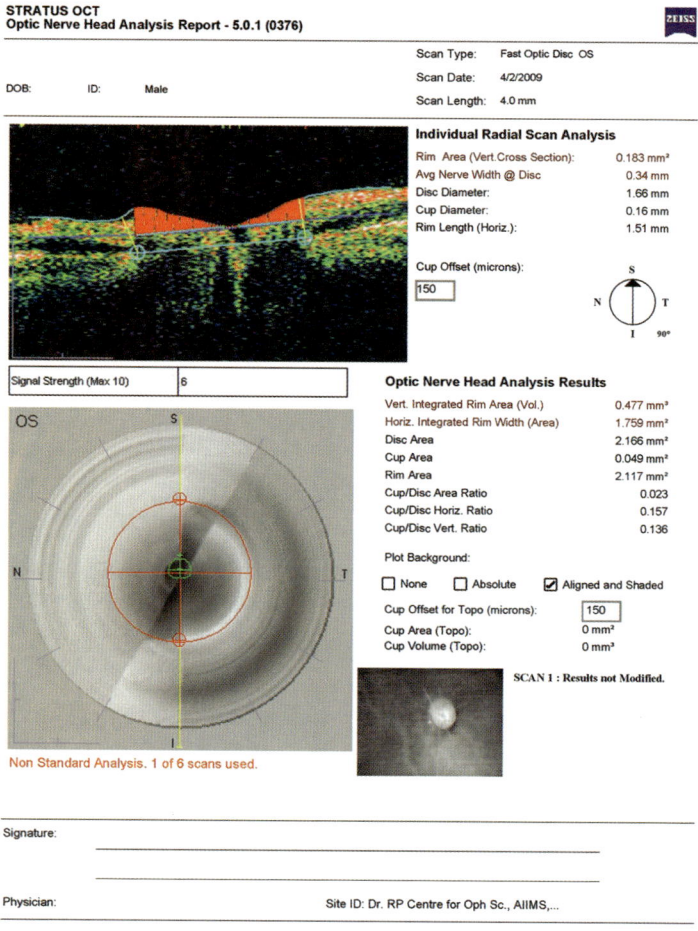

Fig. 4.1 An optic nerve head analysis scan on the Stratus OCT: *Optic Nerve Head Analysis Report* showing a large disc with normal parameters. Note the 150 microns cup offset

this line, various optic nerve head parameters are automatically calculated. These include:
- *Neuro-retinal rim area (mm^2):* Summary of the darker gray neuroretinal rim region shown on top of the RNFL thickness map (Figure 4.5).
- *Cup area (mm^2):* Lighter gray region on RNFL thickness map (Figure 4.5).
- *Disc area:* Refers to the area of the rim plus the area of the cup (mm^2).
- *Average C/D ratio:* Square-root of the ratio of the area of the cup to the area of the disc.
- *Vertical C/D ratio:* Ratio of the cup diameter to the disc diameter in the vertical meridian.

Optic Nerve Head Analysis 27

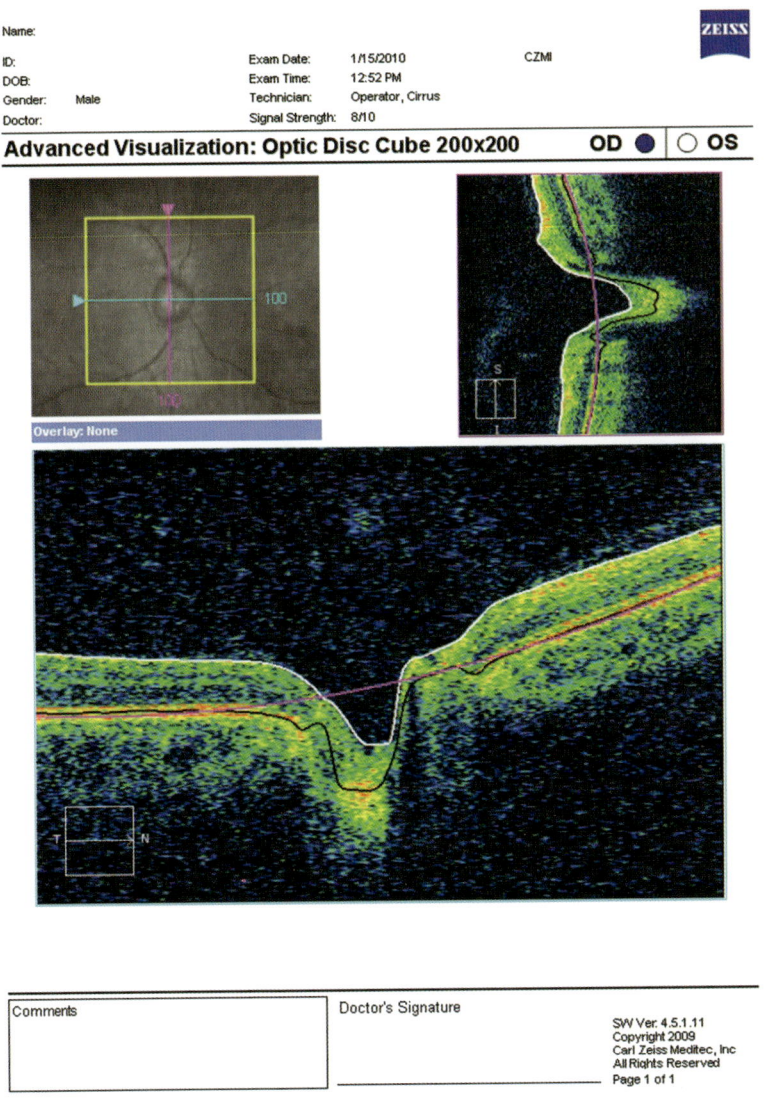

Fig. 4.2 An optic nerve head analysis scan on the Cirrus OCT: *Advanced visualization of the optic disc cube* showing 2D visualization of the optic nerve head

- *Cup volume:* 3D measurement defined as the volume between a plane created by the cup outline at the vitreous interface and the posterior surface of the ONH. Its units are in mm^3.
- *Vertical integrated rim area (Volume):* This estimates the total volume of RFNL tissue in the rim.
- *Horizontal integrated rim width (Area):* This estimates the total rim area.

- *Neuroretinal rim (NRR) thickness:* Thickness (in micrometers) is plotted for left and right eye together in the four quadrants. It may be selected at any line over 360 degree to get NRR data in that aspect. While the ISNT rule forms the basis of glaucoma diagnosis on fundus examination, the same does not hold true for the OCT.[10]

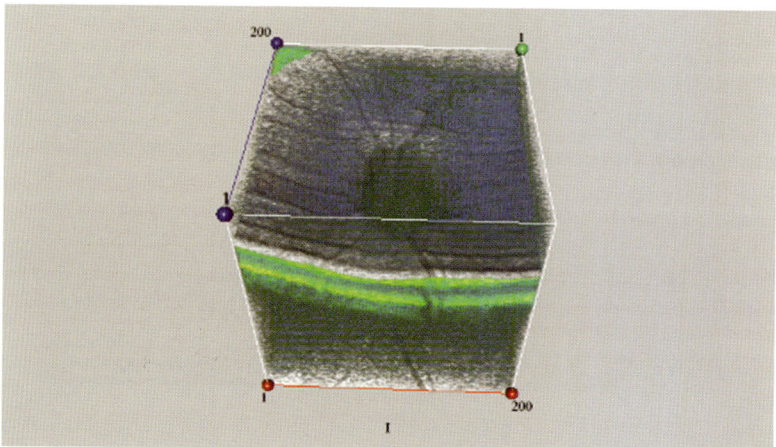

Fig. 4.3A

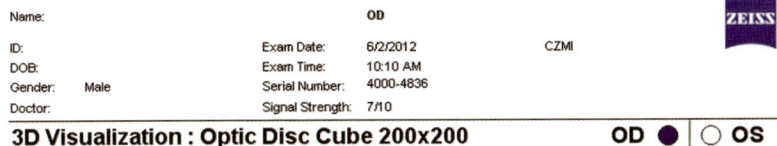

3D Visualization : Optic Disc Cube 200x200   OD ● ○ OS

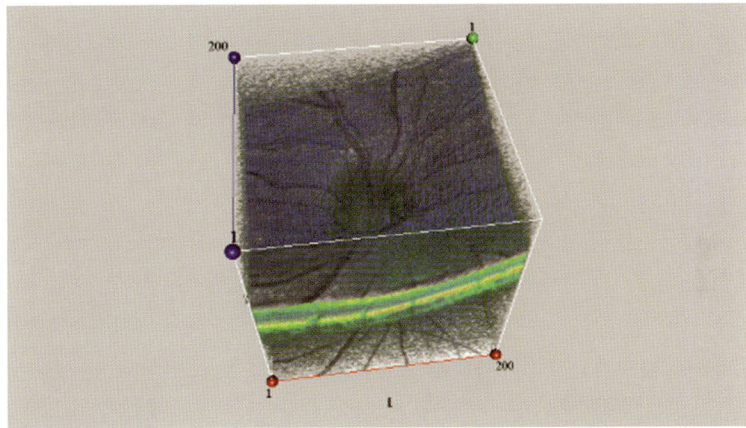

Brightness: 0
Contrast: 100
Threshold: 40
Transparency (%): 0
Use Same Transparency for all Pixels: False

Apply Intensity Filter: False
Intensity Value: 0
Intensity Range: 0
Surface Light Weight: 25
Gradient Step Size: 5
Lighting Enabled: True

Cube cut at Width: 1, 200 Height: 1, 200 A-Scan: 1, 1024

Fig. 4.3B

Figs 4.3A and B A 3D reconstruction of the optic nerve head on the Cirrus OCT showing a shallow cup (A) and a deep cup (B). The cupping usually appears deeper on the OCT as compared to that seen on fundus examination

- *RNFL OU analysis:* The analysis of the retinal nerve fiber layer thickness is available for both eyes and compared with the normative database (described in chapter 5).
- The various optic nerve head parameters quantified are displayed in tables as in Figure 4.6.

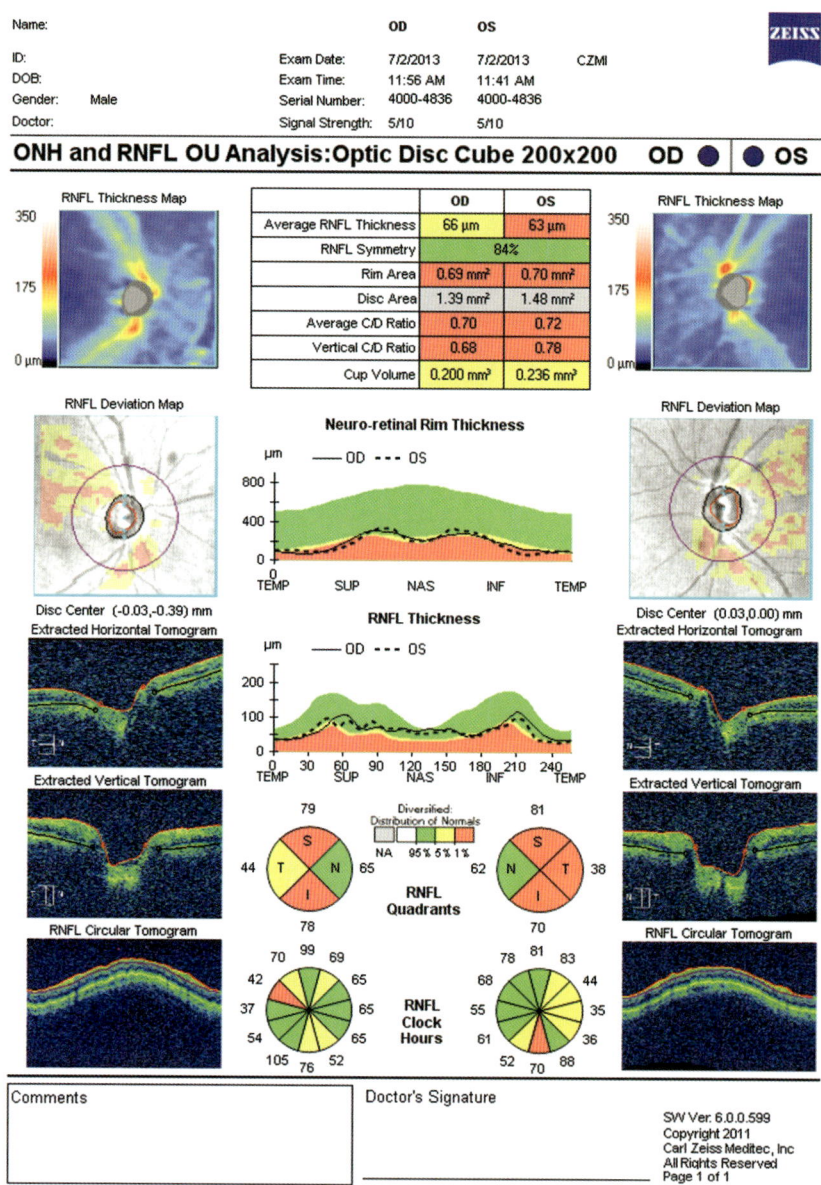

Fig. 4.4 An optic nerve head and RNFL analysis combined scan on the Cirrus OCT. Unlike the Stratus OCT, the optic nerve head objective parameters are displayed and compared with the normative database to enable better glaucoma diagnosis

Optic Nerve Head Analysis

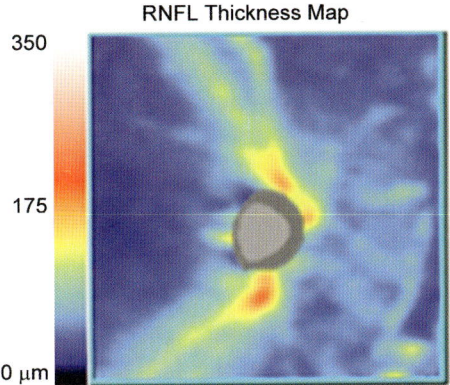

Fig. 4.5 RNFL thickness map from where all disc parameters are derived

Optic nerve head analysis results	
Vert. integrated rim area (vol.)	0.289 mm³
Horiz. integrated rim width (area)	1.881 mm²
Disc area	3.083 mm²
Cup area	1.129 mm²
Rim area	1.954 mm²
Cup/Disc area ratio	0.366
Cup/Disc horiz. ratio	0.606
Cup/Disc vert. ratio	0.604

	OD	Os
Average RNFL thickness	63 µm	77 µm
RNFL symmetry	67%	
Rim area	0.70 mm²	0.92 mm²
Disc area	1.40 mm²	1.12 mm²
Average C/D ratio	0.71	0.43
Vertical C/D ratio	0.97	0.50
Cup volume	0.206 mm³	0.043 mm³

Fig. 4.6 The various optic nerve head parameters studied in the Stratus (Left) and the Cirrus (Right) OCT. Note that the vertical integrated rim area and width are not displayed in the Cirrus OCT ONH printout. Similarly, the horizontal cup disc ratio is also eliminated due to the low relevance for cases of glaucoma. Red coloration of the data indicates that it is normal in less than 1% of the population while a yellow coloration indicates that the likelihood of that data being normal is less than 5%. White and green coloration of data implies that the data is within the limits of the normal population. A pink coloration implies there is no normative data available to compare that parameter

■ CONCLUSION

The optic nerve head scan alone provides us with a significant amount of information with regard to the status of the optic nerve head and includes both subjective and objective reports. These are fairly reproducible and though operator dependent to some extent, they are more useful for follow-up of a patient than diagnosis alone. However for a proper follow-up and prognosis, it is prudent to look at the complete picture and the ever growing role of the RNFL analysis described later.

> **Key Points**
> - Optic nerve head analysis can help diagnose glaucoma.
> - Optic nerve head analysis parameters have a high sensitivity and specificity for glaucoma detection.
> - NRR thickness and rim area/volume are important parameters.
> - ONH analysis should be done cautiously in large discs and myopic eyes.
> - Current OCT machines have normative database for optic nerve head parameters and this enables better diagnosis and follow up.

REFERENCES

1. Mwanza JC, Oakley JD, Budenz DL, Anderson DR. Cirrus Optical Coherence Tomography Normative Database Study Group. Ability of cirrus HD-OCT optic nerve head parameters to discriminate normal from glaucomatous eyes. Ophthalmology. 2011;118:241-8.
2. Kotowski J, Wollstein G, Ishikawa H, Schuman JS. Imaging of the optic nerve and retinal nerve fiber layer: An essential part of glaucoma diagnosis and monitoring. Surv Ophthalmol. 2013.
3. Sung KR, Na JH, Lee Y. Glaucoma diagnostic capabilities of optic nerve head parameters as determined by Cirrus HD optical coherence tomography. Journal of Glaucoma. 2012;21(7):498-504.
4. Onmez FE, Satana B, Altan C, Basarir B, Demirok A. A comparison of optic nerve head topographic measurements by stratus OCT in patients with macrodiscs and normal-sized healthy discs. J Glaucoma. 2013.
5. Na JH, Sung KR, Lee JR, Lee KS, Baek S, Kim HK, Sohn YH. Detection of glaucomatous progression by spectral-domain optical coherence tomography. Ophthalmology. 2013;120(7):1388-95.
6. Lisboa R, Paranhos A Jr, Weinreb RN, Zangwill LM, Leite MT, Medeiros FA. Comparison of different spectral domain OCT scanning protocols for diagnosing preperimetric glaucoma. Invest Ophthalmol Vis Sci. 2013;54(5):3417-25.
7. Sigler EJ, Mascarenhas KG, Tsai JC, Loewen NA. Clinicopathologic correlation of disc and peripapillary region using SD-OCT. Optom Vis Sci. 2013;90(1):84-93.
8. Gardiner SK, Ren R, Yang H, Fortune B, Burgoyne CF, Demirel S. A method to estimate the amount of neuroretinal rim tissue in glaucoma: comparison with current methods for measuring rim area. Am J Ophthalmol. 2013.
9. Mwanza JC, Oakley JD, Budenz DL, Anderson DR. Ability of cirrus HD-OCT optic nerve head parameters to discriminate normal from glaucomatous eyes. Ophthalmology 2011;118:241-8.
10. Hwang YH, Kim YY. Application of the ISNT rule to neuroretinal rim thickness determined using cirrus HD optical coherence tomography. J Glaucoma. 2013.

CHAPTER

5

Retinal Nerve Fiber Analysis

■ INTRODUCTION

The importance of the retinal nerve fiber has been long known. Even prior to the optical coherence tomography (OCT) era, the retinal nerve fiber analysis (RNFL) was carefully examined during ophthalmoscopy to look for defects in all cases of glaucoma. Even in the early phase of development and adaptation of the OCT for clinical use in ophthalmology, the RNFL thickness was a vital output parameter. Today, the ever growing literature on RNFL changes in glaucoma and other ocular and systemic disease is evidence to the fact that this is probably one of the most important parameters derived from the OCT.[1,2]

With regard to glaucoma diagnosis using RNFL as a parameter, it is largely believed that both the *stratus* and the *cirrus* OCT have similar specificity.[3] However literature has also shown that in early glaucoma, cirrus outperforms the startus as also to the GDx-VCC.[4,5] The RNFL measurements in both the OCTs are different as the limits of measurement are the RPE or the IS-OS junction (see differences between the two OCTs in chapter 2). While a formula has been derived to allow interconversion of the RNFL values, it is best to follow-up a patient on either one of the two OCTs.[6]

RNFL is a standard for glaucoma diagnosis and the best OCT parameters include clock hour wise RNFL thickness maps (better than deviation maps), RNFL thickness in the inferior, superior, and 1 o'clock hour sector; the cup area; and the vertical integrated rim area of the optic disc.[7-10] Addition of the rim area to the RNFL parameters has shown to greatly enhance the sensitivity and specificity of diagnosis.[11]

RNFL measurements may get affected by various parameters. These include, decreasing angle between the inferotemporal and superotemporal veins, thinner central corneal thickness, and increasing axial length each of which appears to increase the separation of the RNFL peaks.[12] Disc size has an inverse effect on the average, superior and temporal RNFL.[13]

A major advantage of the SD-OCT and RNFL measurements is that they have a good long-term reproducibility and are reliable measures.[14]

The newer aspect emerging in the field of RNFL analysis using the OCT is a concept of nerve fiber reflectance which holds potential in early diagnostic ability of glaucoma particularly in cases of early glaucoma.[15]

Cirrus OCT has been shown to have a sensitivity and specificity for average RNFL as 83% and 88%, respectively (p<0.05), and 65% and 100%, respectively (p<0.01) for detecting glaucoma.[16] Stratus OCT has been shown to have a sensitivity and specificity for average RNFL as 80% and 94%, respectively (p<0.05), and 61% and 100%, respectively (p<0.01) for diagnosis of glaucoma.[16]

■ METHOD

RNFL image acquisition may be done in the absence of pupillary dilatation (minimum diameter: 3 mm) in the newer OCT machines, however a dilated pupil eases the testing process and avoids any inadvertent errors. The patient is asked to fixate on an internal fixation target in the same manner as for the optic nerve head scan (described previously) and this helps in improving reproducibility of the image. A circular (or rarely linear) aspect of the RNFL is usually tested and the acquisition is done around either the optic disc or the macula. From these images, a real-time two-dimensional tomographic image is constructed.

■ TECHNOLOGY

The first reflection captured in the 2D image is the vitreous-internal limiting membrane interface. The second highly reflective interface posterior to this is the retinal pigment epithelium-photoreceptor interface. These define the boundaries of retina. Mean RNFL thickness is calculated using the inbuilt RNFL thickness average analysis protocol. The boundaries of RNFL are defined by first determining the thickness of the neuro-sensory retina. The location of posterior boundary of RNFL is determined by evaluating each A-scan for a threshold value chosen to be 15 dB greater than the filtered maximum reflectivity of the adjacent retina. Average measurements are given for twelve 30° sectors. The depth values of the scans are independent of the optical dimensions of the eye, and no reference plane is required unlike the HRT.

■ ANALYSIS

The OCT usually automatically selects the acquisition circle for RNFL measurement while performing the ONH scan in the newer machine and the user does not need to make any modifications. During an RNFL scan, the disc is automatically delineated and even though this can be done by the operator, the automated system is shown to work just as well.[17] Furthermore, the acquisition circle can be changed into a grid and that has shown better sensitivity.[18]

While everything is automated, it is of interest to know that the OCT offers a variety of RNFL thickness measurement and analysis protocols:
1. *RNFL thickness protocol (3.4 mm):* Acquires a scan with radius 1.73 mm, centered on the optic disc.

2. *Fast RNFL thickness protocol (3.4 mm):* Acquires three fast circular scans. This is time efficient scan alignment and placement is required only once. However, it may be slightly less accurate and valid for repeat measurements.
3. *Proportional circle:* This protocol allows measurement of RNFLT around the optic disc along a circular scan, the size of which can be tailored taking into account the size of optic nerve head.
4. *Concentric 3 rings:* This protocol enables us to measure RNFLT along three equally placed default circular scans of 0.9 mm, 1.81 mm and 2.71 mm radii. However, the scan radius can be altered according to the need.
5. *RNFL map:* This protocol comprises of six circular scans of 1.44 mm, 1.69 mm, 1.90 mm, 2.25 mm, 2.73 mm, and 3.40 mm radii. This gives an overlay view of the RNFLT, around the peripapillary area. Retinal nerve fiber layer measurement with a circular scan of 1.34 mm radius, centered on the optic nerve head has been shown to have a maximum reproducibility.
6. *RNFL thickness (2.27 times the disc size):* This circular RNFLT scan size is 2.27 times the radius of the optic nerve head. This helps in measuring RNFL with good reproducibility.

However, the standard RNFL measurement on the cirrus OCT, when an optic nerve head 200 × 200 scan is selected is done is at 3.54 mm diameter circle around disc which for an average sized disc is 2.27 times the disc size and is usually adequate for diagnosis and follow-up.

■ INTERPRETATION

Interpreting the RNFL OU Analysis Screen

The typical RNFL printout obtained in the cirrus OCT is depicted in Figure 5.1. The earlier stratus OCT had a different output printed which is depicted in Figure 5.2. The various components of these printouts are discussed below:
1. *RNFL deviation map:* This image is a sum of the reflectivity in each A-scan, and illustrates the anatomy scanned. The image is overlaid on with deviation of the RNFL thickness from normal which is depicted in the form of red superpixels or yellow superpixels. Also shown in purple is the calculation circle used for RNFL thickness measurements. The black line depicts the optic disc outline while the red line demarcates the cup (Figure 5.3). It is important to note that unlike the RNFL deviation map which shows RNFL thinning as red and yellow, the RNFL thickness map shows the area of RNFL thinning as blue (Figure 5.4).
2. *Optic nerve head gray scale:* The stratus OCT, in addition, also showed a gray scale representation of the cup and neuroretinal rim. The interior of the cup is shown in light gray while the NRR is shown in dark gray. The outer boundary of the NRR corresponds to the disc boundary.
3. *Extracted B-scan of optic nerve head:* The cup and disc boundaries are illustrated is on the extracted OCT B-scan (Figure 5.5). The B-scan data is extracted from a 4 mm radial line that cuts through the center of the disc. The line is shown in turquoise on top of the OCT fundus overlaid on the gray scale fundus image. The segmented RPE layer is shown on the B-scan as a

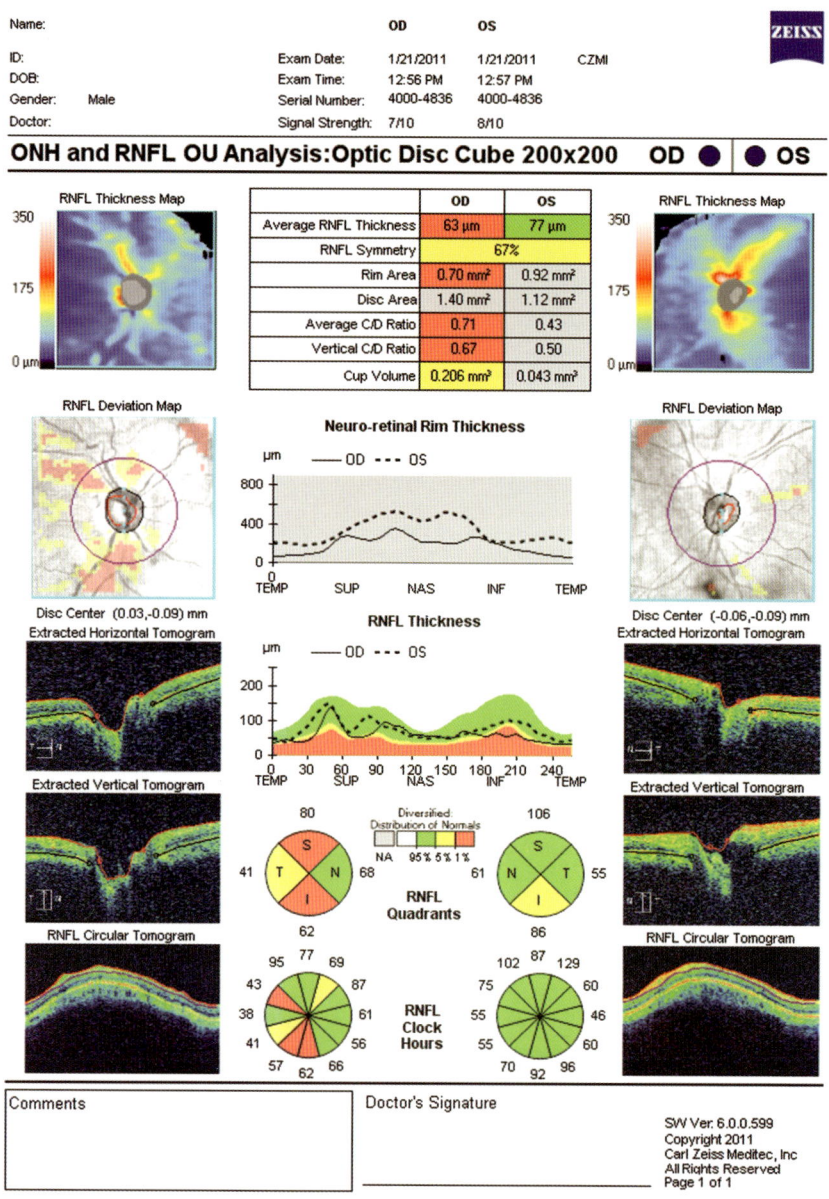

Fig. 5.1 The typical printout generated in cirrus OCT after RNFL analysis

black line, and the disc boundaries are shown in this 2D picture as black markers. The segmented ILM is shown as a red line, and the cup boundaries are shown in this 2D picture as red markers. The radial line can be changed in steps of 5° to see the B scan along different axis of the disc (Figure 5.6).

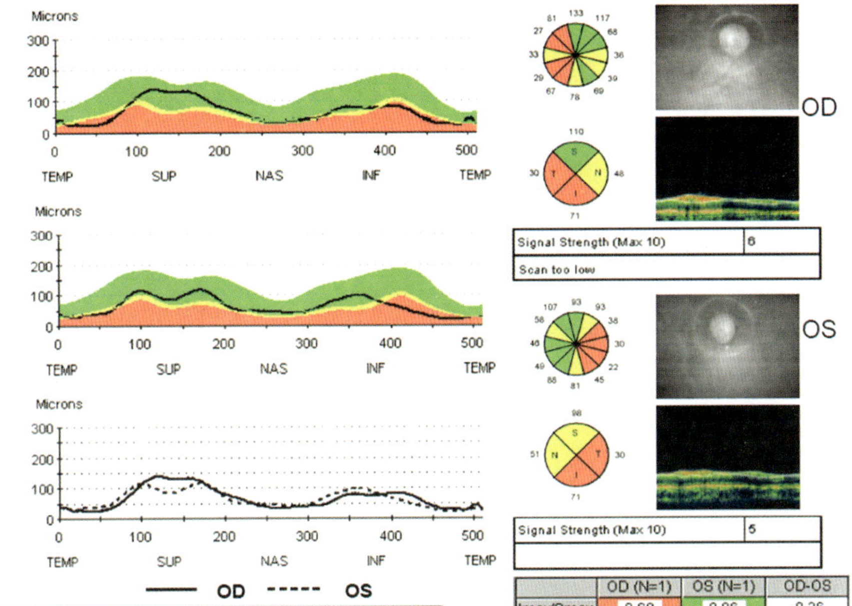

Retinal Nerve Fiber Analysis

Fig. 5.2 The typical printout generated in a stratus OCT after RNFL analysis

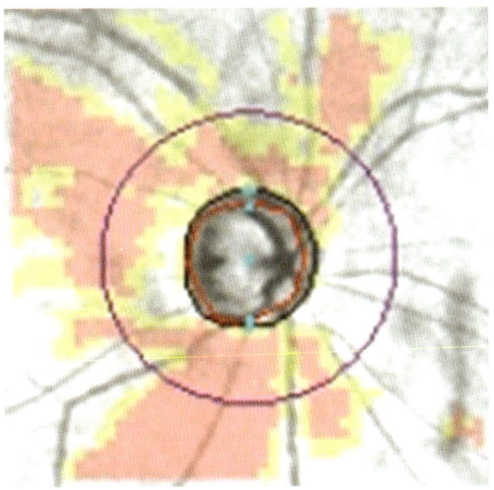

Fig. 5.3 RNFL deviation map

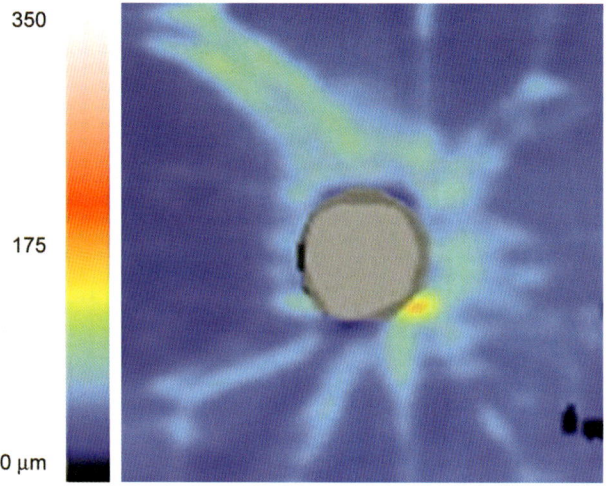

Fig. 5.4 RNFL Thickness map

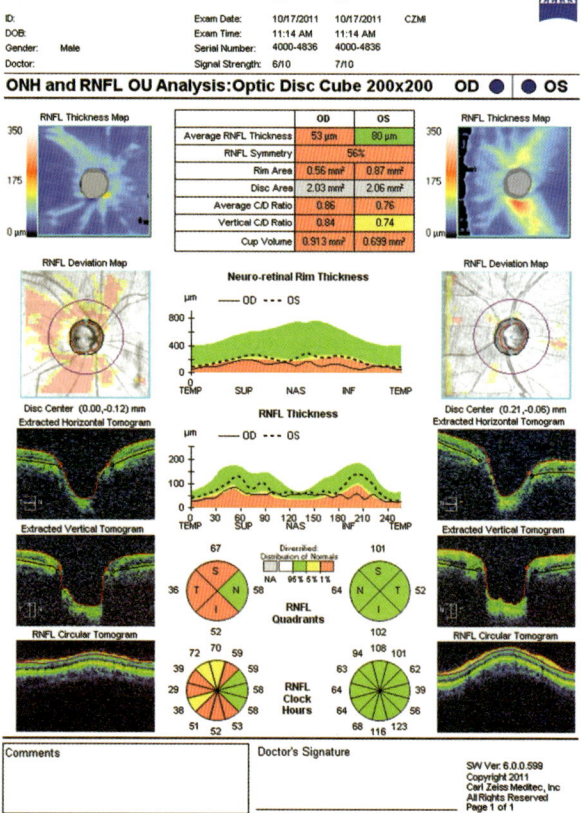

Fig. 5.5 A typical printout of an RNFL analysis. Note the extracted B-scans on the bottom left and right columns of the output image. These form the mainstay for RNFL measurement and should not be truncated at the top or bottom while acquiring the image

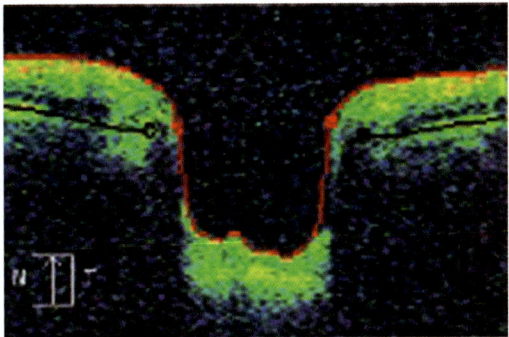

Fig. 5.6 Extracted vertical tomogram

RNFL circular tomogram

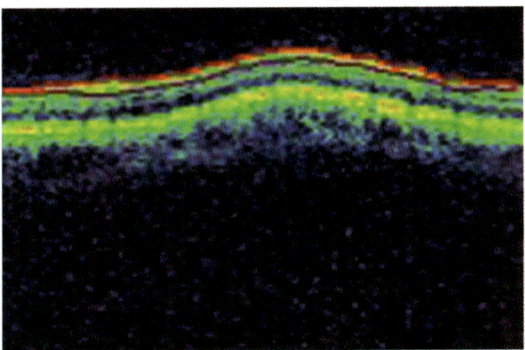

Fig. 5.7 Extracted B-scan showing RNFL

4. *Extracted B scan of RNFL:* An extracted B-scan of the RNFL thickness along the 3.54 mm circle is also depicted (Figure 5.7).
5. *RNFL thickness map:* An RNFL deviation map is also displayed and as always represented; warm colors are thicker while cool colors are thinner (Figures 5.4 and 5.8). This makes it easy to recognize focal or wedge-shaped RNFL defects at a glance.
6. *Average RNFL thickness (Data table):* Mean RNFL thickness is calculated using the inbuilt RNFL thickness average analysis protocol. Layer-seeking algorithms find the RNFL inner (anterior) boundary and RNFL outer (posterior) boundary for the entire cube, excepting the optic disc. The system extracts from the data cube 256 A-scan samples along the path of the calculation circle that together comprise the RNFL scan image displayed. Based on the RNFL layer boundaries in the extracted circle scan image, the system calculates the RNFL thickness at each point along the calculation circle. The thickness data is plotted in the right and left eye thickness graphs and the symmetry comparison graph. The earlier stratus OCT also displayed S_{avg} and I_{avg} (Superior average and inferior RNFL average) and S_{max} and I_{max} (Superior maximum and inferior maximum RNFL thickness) and Max-Min difference (Figure 5.9).

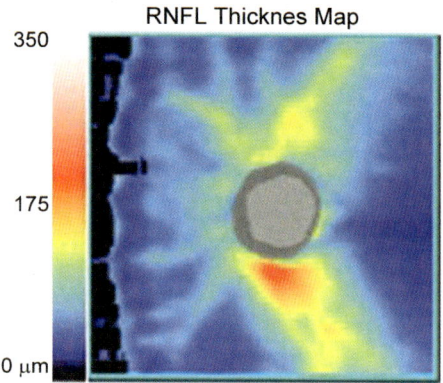

Fig. 5.8 RNFL thickness map showing bifid superior RNFL

	OD (N=1)	OS (N=1)	OD-OS
	0.60	0.86	-0.26
Smax/Imax	1.66	1.16	0.49
Smax/Tavg	4.61	3.88	0.73
Imax/Tavg	2.78	3.34	-0.56
Smax/Navg	2.90	2.29	0.61
Max-Min	114.00	96.00	18.00
Smax	138.00	117.00	21.00
Imax	83.00	100.00	-17.00
Savg	110.00	98.00	12.00
Iavg	71.00	71.00	0.00
Avg. Thick	64.79	62.49	2.30

	OD	OS
Average RNFL thickness	63μm	77μm
RNFL symmetry	67%	
Rim area	0.70 mm^2	0.92 mm^2
Disc area	1.40 mm^2	1.12 mm^2
Average C/D ratio	0.71	0.43
Vertical C/D ratio	0.67	0.50
Cup volume	0.206 mm^3	0.043 mm^3

Fig. 5.9 RNFL thickness data table

7. *TSNIT neuroretinal rim thickness profile:* TSNIT stands for temporal, superior, nasal, inferior and temporal RNFL. This displays NRR thickness at each A-scan location along the calculation circle and displays both eye results on the same graph to enable a visual comparison (Figure 5.10).
8. *TSNIT RNFL thickness profile:* This displays RNFL thickness at each A-scan location along the calculation circle and includes as a backdrop the white-green-yellow-red color code based on the age-matched RNFL normative data (Figure 5.11). The profile shows left and right eye RNFL thickness together, to enable comparison of symmetry in specific regions. Drag the blue vertical line in the OU profile to select the current A-scan sample from among the 256 samples. A similar vertical blue line tracks the current sample in the RNFL circle scan image.

The ONH and RNFL OU analysis supports the clinician in identifying areas of the RNFL that may be of clinical concern by comparing the measured RNFL thickness to age-matched data. Normative data that is age-matched to the patient appears when you perform the ONH and RNFL OU analysis on patients at least 19 years old since there is no data of patients younger than that in the machines database. Also there may be racial differences in the normative data and it is more useful to have your machine updated with the normative database from the race/country you are evaluating/treating.

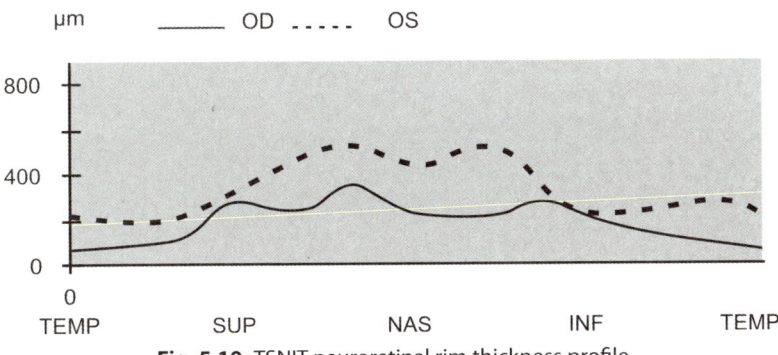

Fig. 5.10 TSNIT neuroretinal rim thickness profile

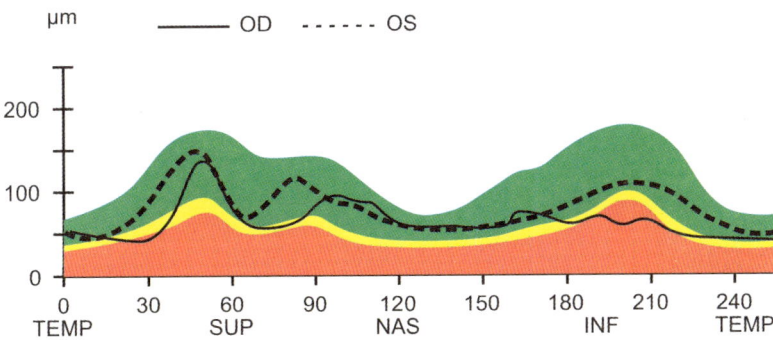

Fig. 5.11 TSNIT RNFL thickness profile

The RNFL normative database uses a white-green-yellow-red color code, as seen in the legend at left, to indicate the normal distribution percentiles. The color code applies to each particular A-scan location in the TSNIT thickness graphs, to the quadrant, clock hour and whole-circle averages, and to the OD and OS columns of the data table. Among same-age individuals in the normal population, the percentiles apply to each particular RNFL thickness measurement along the calculation circle as follows:

- The thinnest 1% of measurements falls in the red area. Measurements in red are considered outside normal limits (red < 1%, this means that there is a less than 1% chance that this particular value belongs to within normal limits; outside normal limits).
- The thinnest 5% of measurements fall in the yellow area or below (1% >yellow < 5%, suspect).
- Ninety percent of measurements fall in the green area (5% ≤ green ≤ 95%; normal).
- The thickest 5% of measurements fall in the white area (white > 95%; normal).

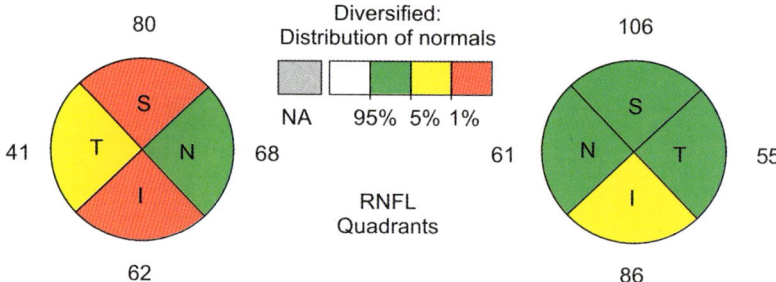

Fig. 5.12 Quadrantic RNFL thickness maps

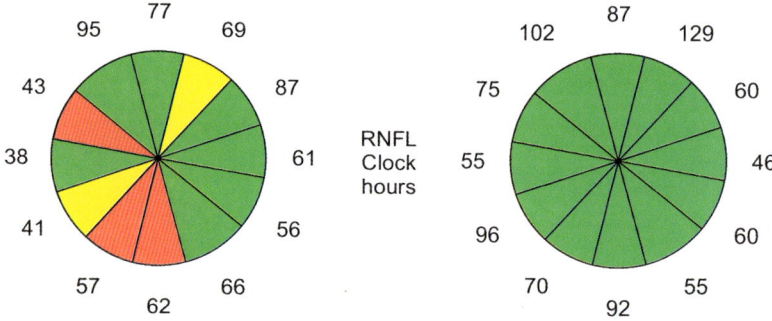

Fig. 5.13 Clock hour RNFL thickness maps

9. *Quadrantic RNFL thickness maps:* These depict the RNFL thickness in the four quadrants and color code them in concordance with the deviation from the normative database (described here) (Figure 5.12).
10. *Clock hour RNFL thickness maps:* These depict the RNFL thickness in each clockhour and also use a color coding in concordance to the deviation from the normative data (Figure 5.13).

■ CONCLUSION

The retinal nerve fiber layer is an important parameter evaluated in both a qualitative and quantitative manner by the OCT. It is of importance in evlauting the extent of damage due to glaucoma as well for follow-up of glaucoma cases. It is of particular importance in pre-perimetric cases of glaucoma where the visual fields may be normal despite thinning on the OCT. The few cases following this chapter will give a glimpse at some of the situations encountered where RNFL thickness measurement may be useful.

Key Points

- RNFL is the most important parameter studied for glaucoma analysis
- RNFL change detection carries a high sensitivity and specificity for glaucoma diagnosis
- Both qualitative and quantitative analysis should be interpreted
- RNFL quadrant analysis and TSNIT plots form the bottom-line for glaucoma detection
- Always clinically correlate the OCT findings as a large number of non-glaucomatous disorders can result in RNFL changes and a misdiagnosis.

CASES

CASE 1: STRUCTURE CORRELATES WITH FUNCTION

This is an elderly male with primary angle closure glaucoma with repeated attacks in the right eye. The fundus photograph shows an increased cupping in the right eye with thinning of the neuroretinal rim primarily in the inferior quadrant with apparently well preserved RNFL in the other quadrants. However, the OCT shows significant thinning of the RNFL in the superior, inferior and the temporal quadrant which corresponds with the superior arcuate field defect having a clear nasal step. The left eye appears normal. Changes on the OCT correlate with functional change.

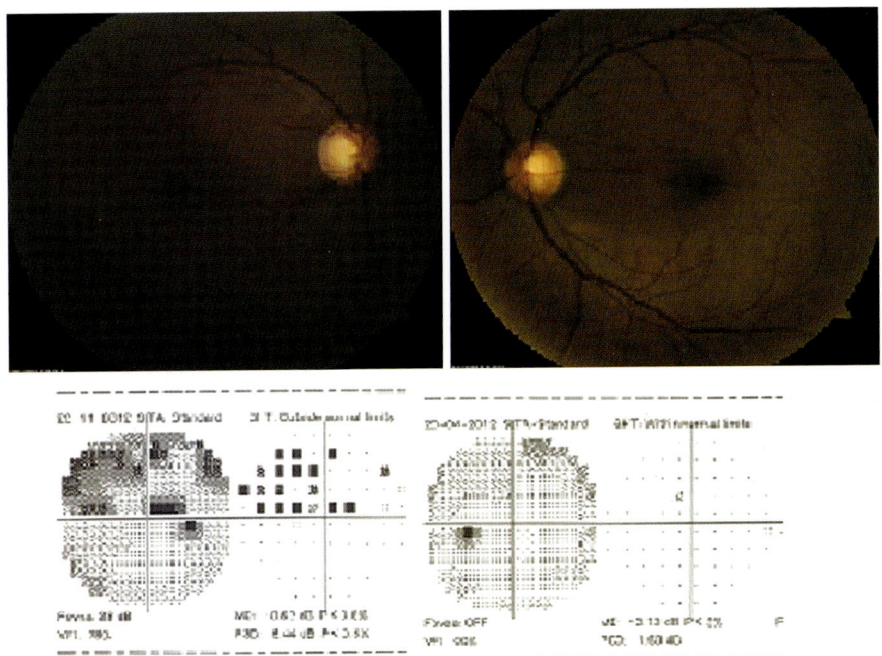

Optical Coherence Tomography in Current Glaucoma Practice

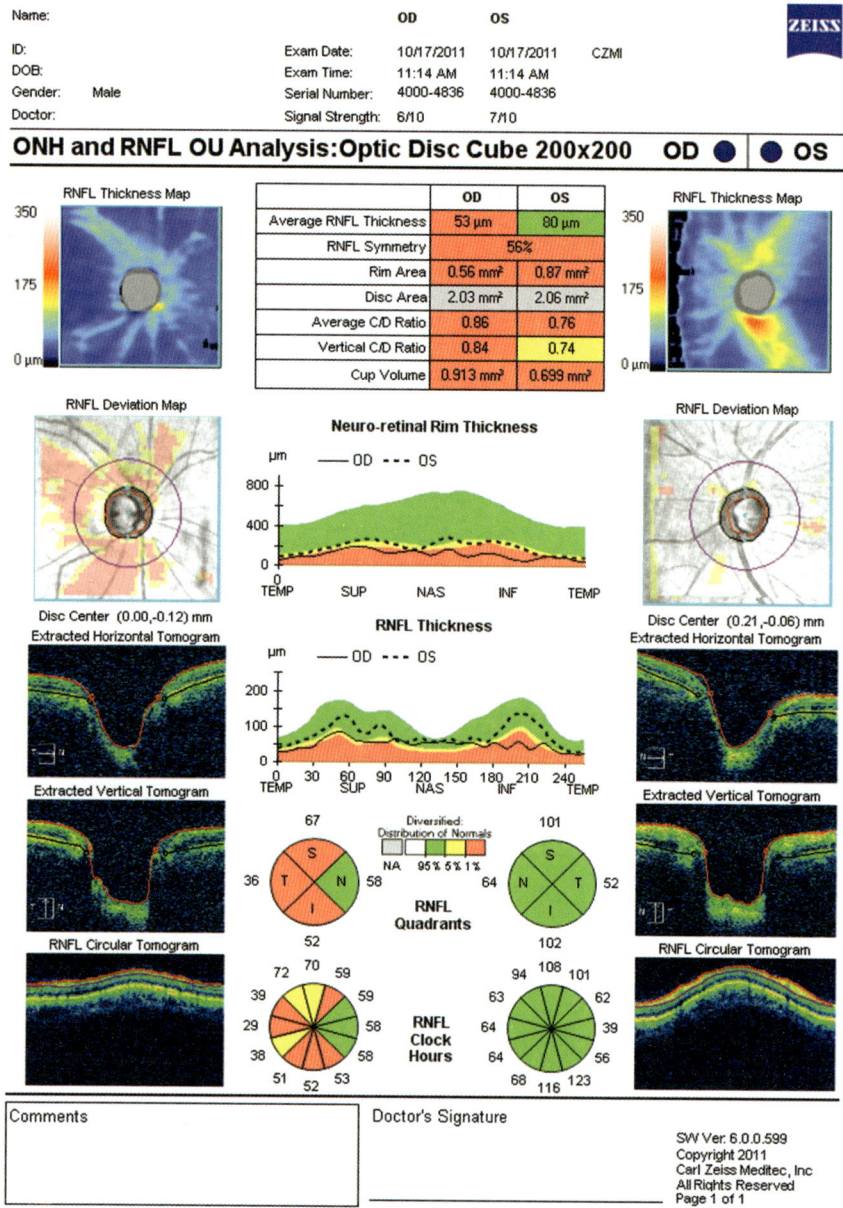

OCT shows thinning of the RNFL of the right eye particularly in the inferior and temporal quadrants. Note the thinning seen on the RNFL deviation map and the clock hour display. Left eye OCT is within normal limits.

CASE 2: PREPERIMETRIC AND PERIMETRIC GLAUCOMA

This is a middle aged patient with a typical glaucomatous visual field defect in the right eye and a near normal field in the left eye. Notice the myopic fundus with large discs and moderate cupping in both the eyes. The OCT is showing a thinning of the superotemporal RNFL in the right eye which is corresponding fairly well with the field defect and the superotemporal notching on the fundus. The left eye OCT is showing some thinning of the RNFL in the inferior and inferotemporal quadrants which does not directly correspond with any field defect. This is a classic situation of pre-perimetric glaucoma in the left eye where the structural changes are evident before the functional changes. Note how the more severe RNFL loss in the right eye has finally led to visual field defect and similarly, further thinning of the RNFL in the left eye would probably eventually lead to a visual field defect.

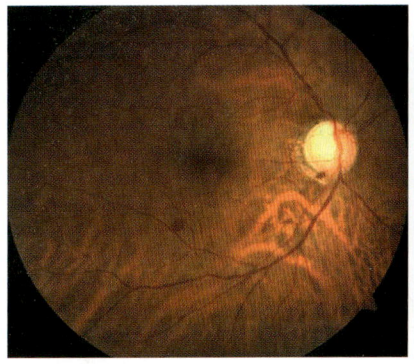
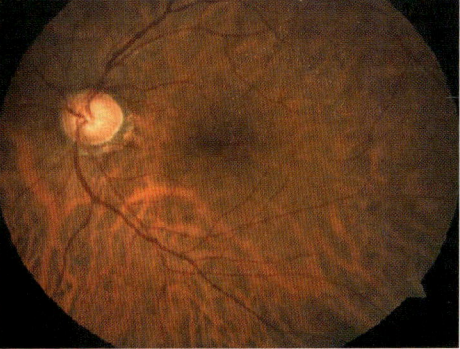
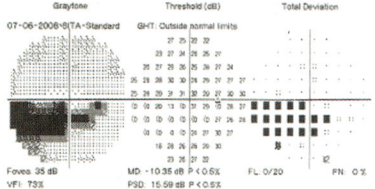

Optical Coherence Tomography in Current Glaucoma Practice

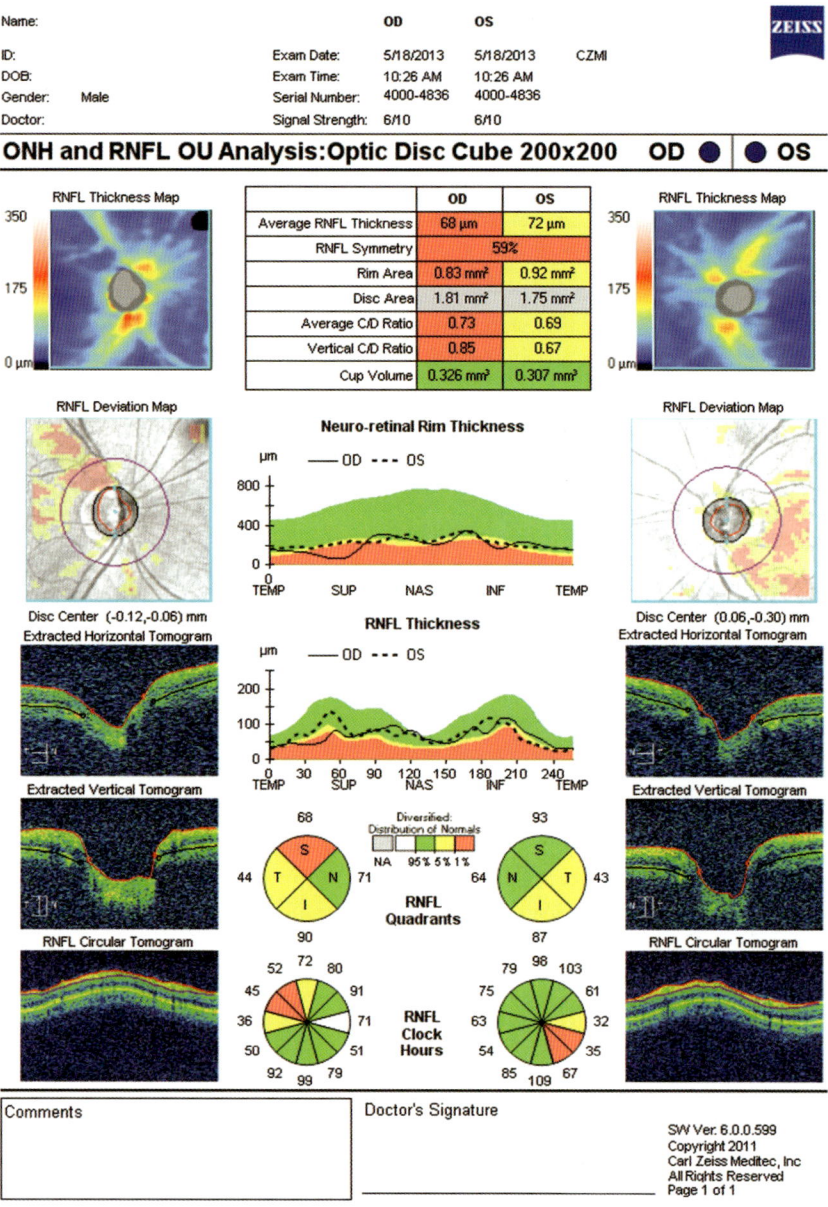

OCT is showing thinning of the RNFL of both eyes with greater thinning in the right eye as compared to the left. In the right eye, there is maximum RNFL loss in the superior quadrant followed by a moderate loss in the temporal and inferior quadrant. The left eye is showing only moderate RNFL thinning in the temporal and inferior quadrant. Clock hour display further demonstrates that there is predominantly in the 10-11 O'clock hour of right eye and the 4-5 O'clock hour of the left eye.

CASE 3: STRUCTURAL CHANGE CORRELATES WITH FUNCTION (IGNORE ARTIFACTS)

This is the case of a middle-aged male patient having secondary glaucoma in the left eye after a prior history of trauma. Notice how the left eye visual fields are showing a superior arcuate scotoma. This field change is corresponding well with the inferior thinning noticed on the RNFL scan of the left eye on the OCT printout. Note the deep blue color of a inferior wedge-shaped defect on the RNFL thickness map and the multiple red superpixels on the RNFL deviation map suggestive of thinning in that region. Also notice an area of sharply demarcated localized thinning in the inferior region of the right eye. This is artifactual due to the presence of a localized pathology such as a pigment epithelial defect in that region.[19]

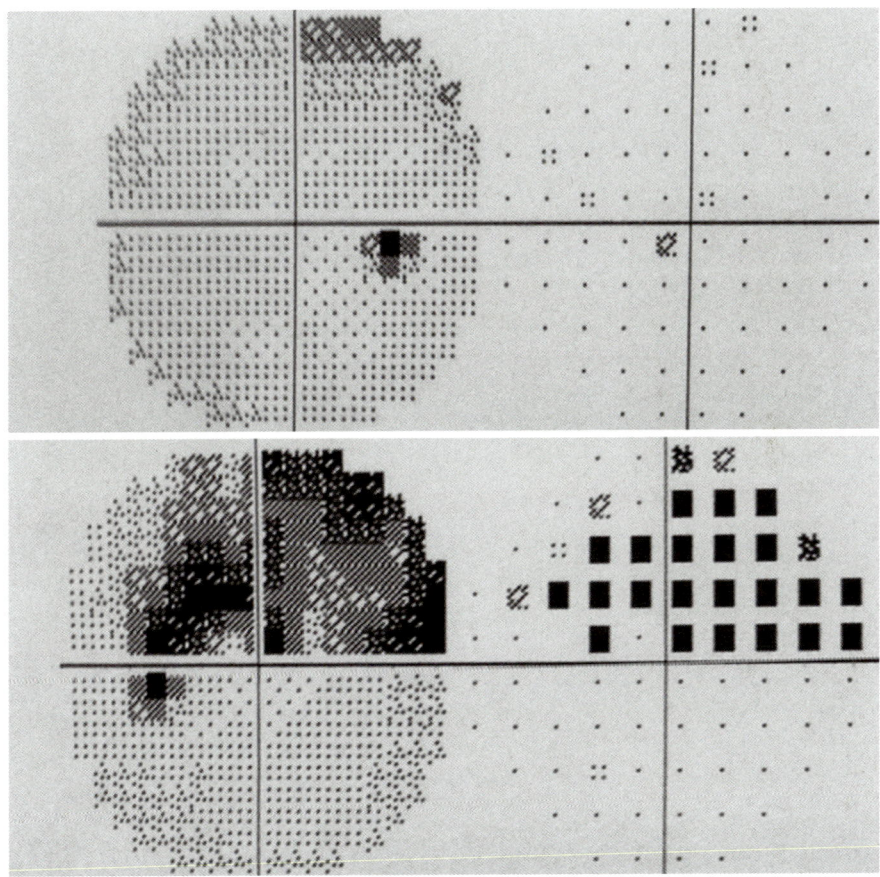

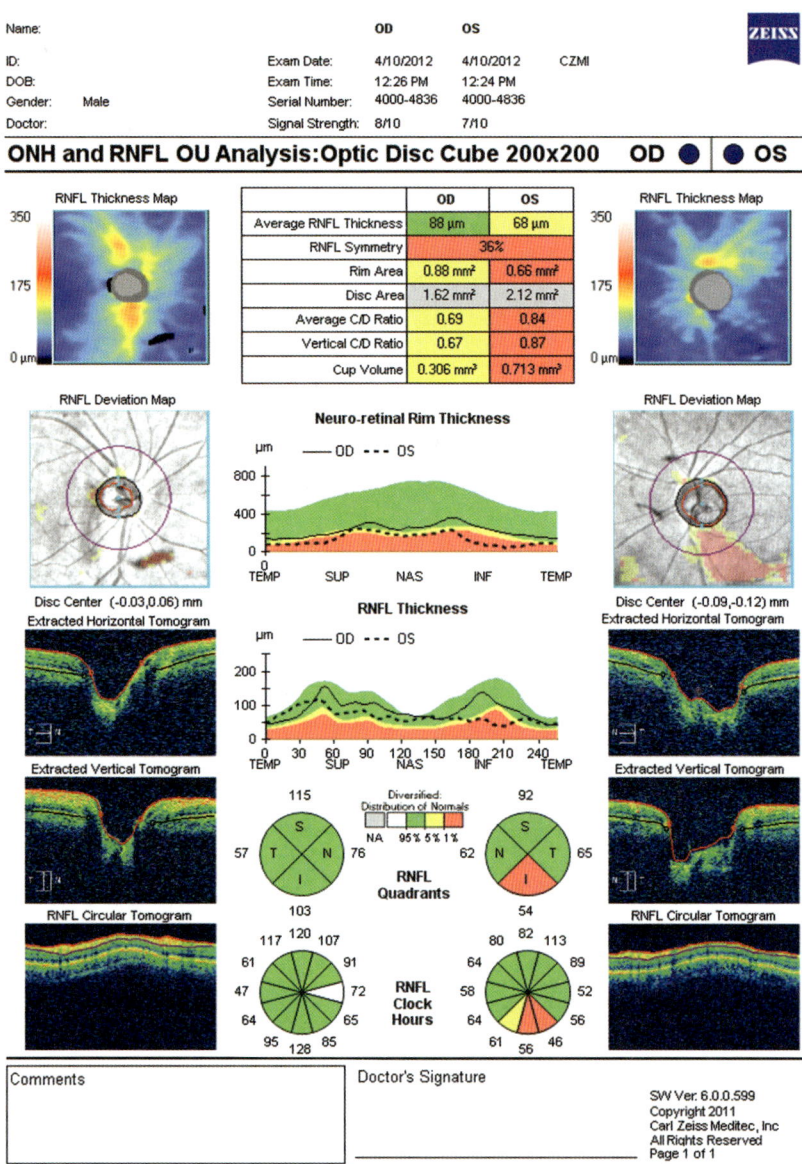

OCT is showing thinning of the RNFL of the inferior quadrant of the left eye. However, there are changes on OCT in both eyes as is evident in the optic nerve head table. This, however, was due to asymmetry in the two disc appearances. Notice a sharply demarcated wedge defect in the RNFL thickness and deviation profiles secondary to a pigment epithelial defect. This is an artifact and should be ignored.

CASE 4: RNFL THINNING COULD BE DUE TO NONGLAUCOMATOUS CAUSES

This is a patient having an asymmetrical pituitary adenoma with visual field loss of both eyes. Notice a bitemporal hemianopia seen on the Goldmann kinetic perimetry with fields in the right eye being worse affected than the left eye. Notice the RNFL analysis done on the stratus OCT showing severe thinning of the RNFL of the right eye with relative preservation of the RNFL of the left eye. This case exemplifies that there could be non-glaucomatous causes for RNFL loss and it must be interpreted with a clinical correlate.

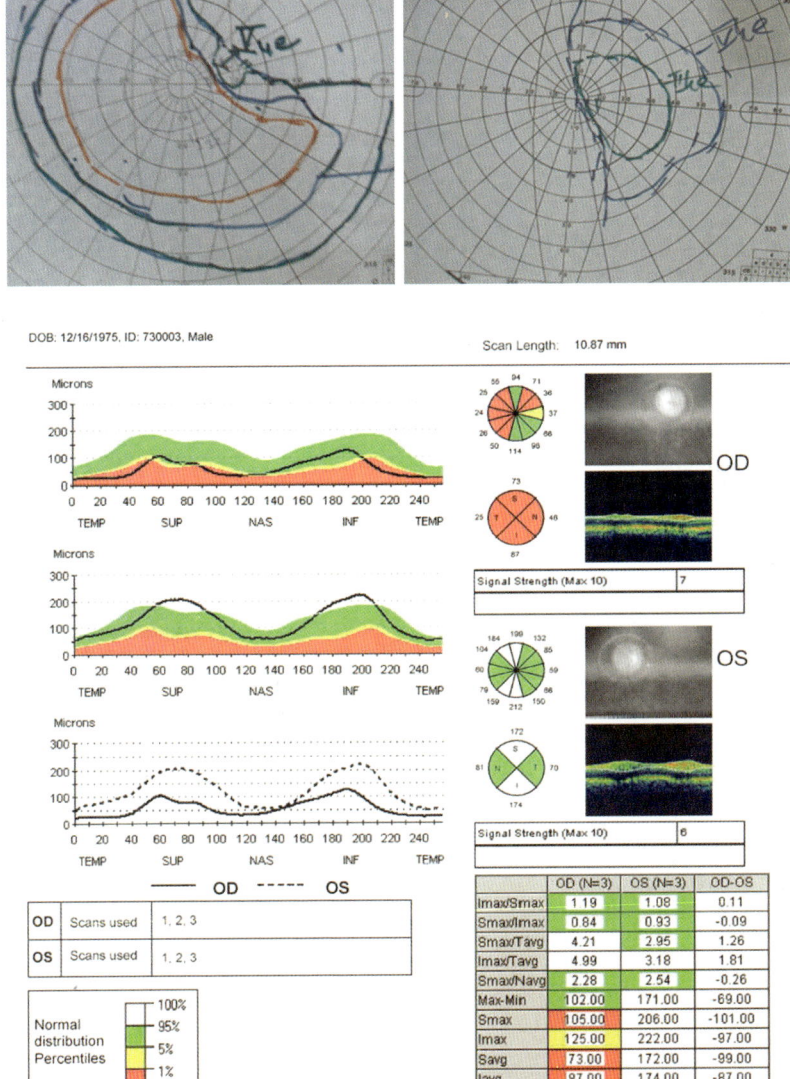

■ CASE 5: UNDERSTANDING THE TSNIT RNFL PROFILE

This case is demonstrating a glaucomatous neuropathy with thinning of the inferior neuroretinal rim and RNFL. Note how the usual double humped TSNIT RNFL profile is blunted at the inferior location and now appears more as a single peaked curve at the superior location. The second OCT which follows shows an even more advanced glaucoma with severe thinning of both the inferior and the superior RNFL. This leads to blunting of both the peaks of the TSNIT profile.

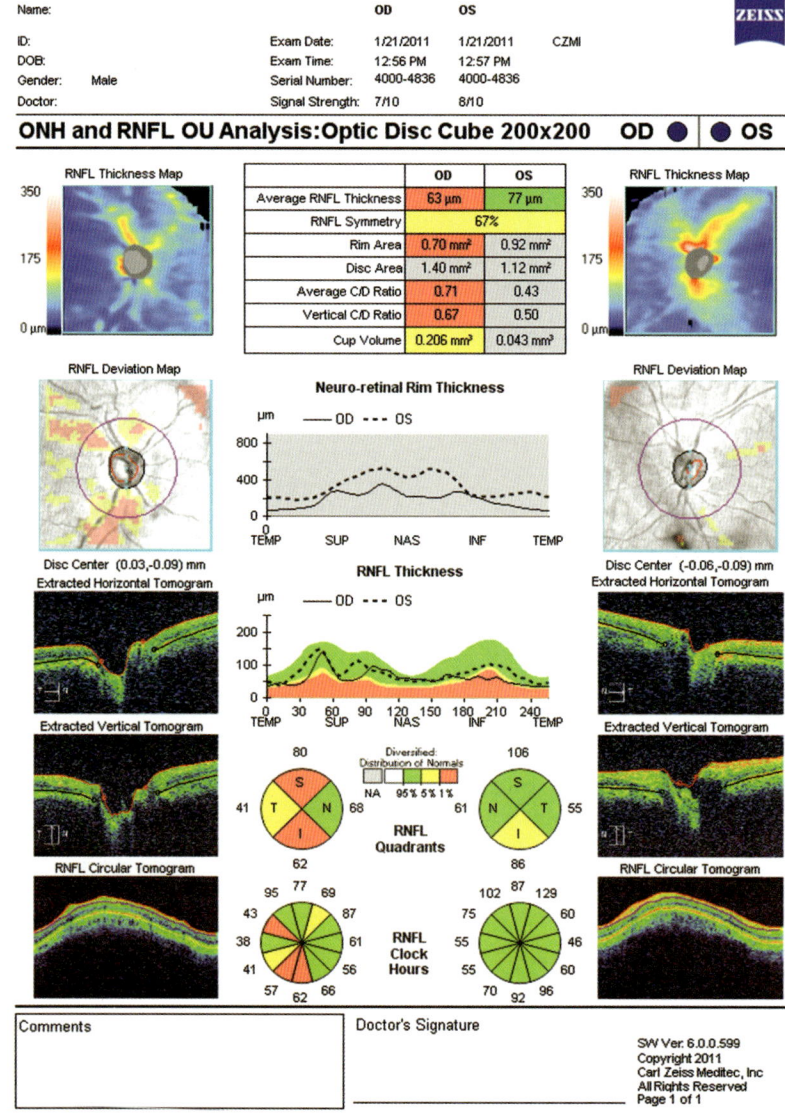

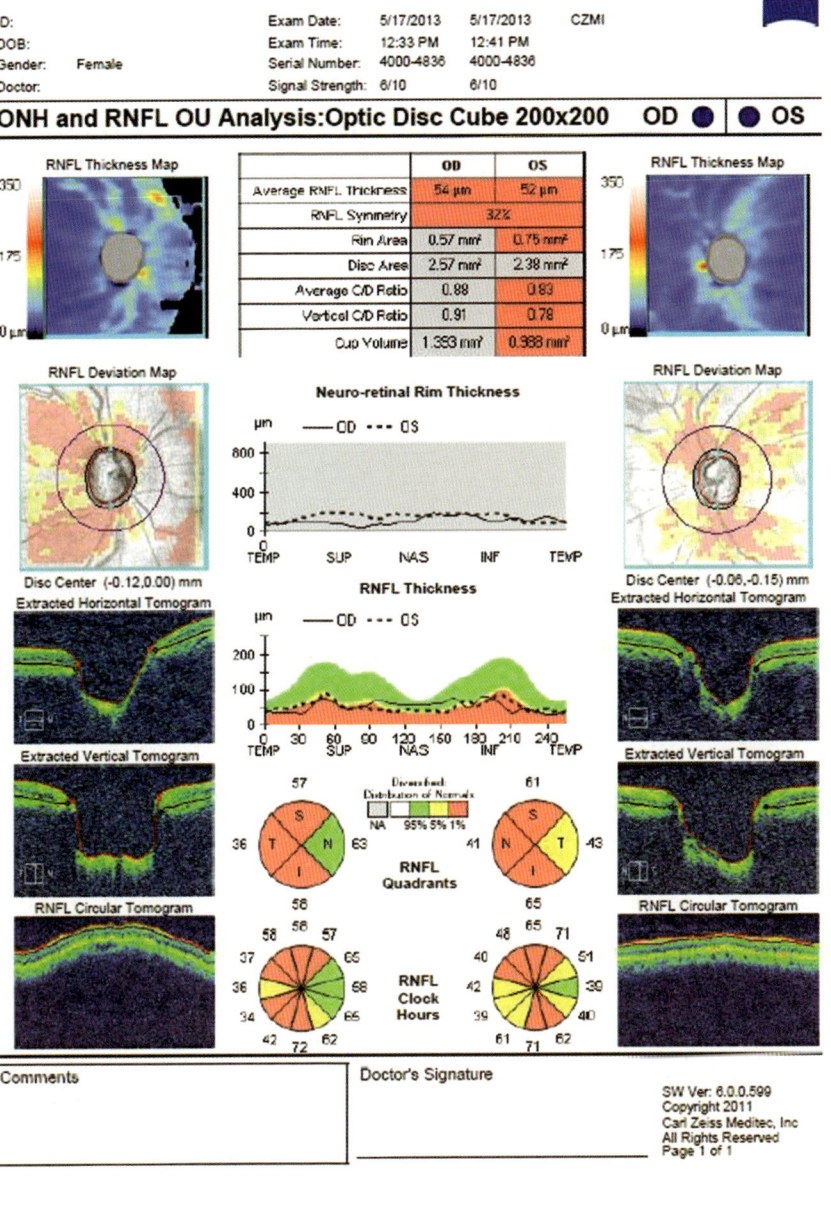

CASE 6: QUALITATIVE ANALYSIS

The printout below shows only the RNFL deviation map and the TSNIT profiles. This is an entirely qualitative method of quickly evaluating the RNFL in a patient. It is also evident from this printout that the RNFL thickness TSNIT profile of both the eyes may be separately displayed instead of the standard single overlaid display. This printout exemplifies that the OCT may be programmed to give only the specific analysis desired by the examinee.

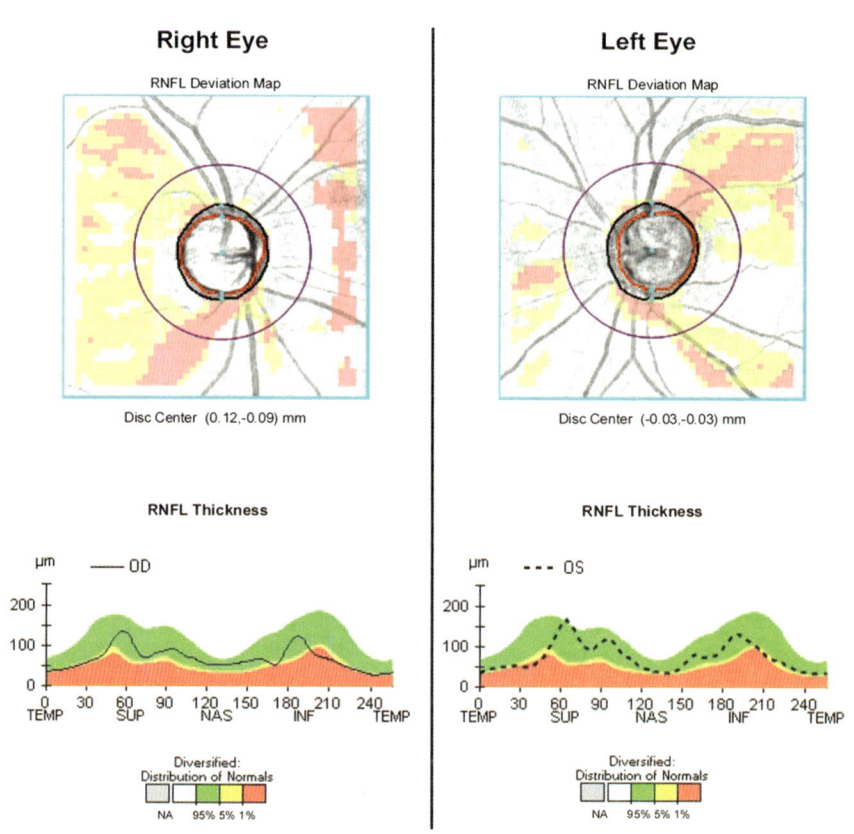

REFERENCES

1. Kotowski J, Wollstein G, Ishikawa H, Schuman JS. Imaging of the optic nerve and retinal nerve fiber layer: An essential part of glaucoma diagnosis and monitoring. Surv Ophthalmol. 2013.
2. Grewal DS, Tanna AP. Diagnosis of glaucoma and detection of glaucoma progression using spectral domain optical coherence tomography. Curr Opin Ophthalmol. 2013;24(2):150-61.
3. Chen HY, Chang YC, Wang IJ, Chen WC. Comparison of glaucoma diagnoses using Stratus and Cirrus optical coherence tomography in differentglaucoma types in a Chinese population. J Glaucoma. 2013;22(8):638-46.
4. Kim SY, Park HY, Park CK. The effects of peripapillary atrophy on the diagnostic ability of Stratus and Cirrus OCT in the analysis of optic nerve head parameters and disc size. Invest Ophthalmol Vis Sci. 2012;53(8):4475-84.
5. Na JH, Lee KS, Lee JR, Lee Y, Kook MS. The glaucoma detection capability of spectral-domain OCT and GDx-VCC deviation maps in early glaucomapatients with localized visual field defects. Graefes Arch ClinExpOphthalmol. 2013;251(10):2371-82.
6. Lee KH, Kang MG, Lim H, Kim CY, Kim NR. A formula to predict spectral domain optical coherence tomography (OCT) retinal nerve fiber layer measurements based on time domain OCT measurements. Korean J Ophthalmol. 2012;26(5):369-77.
7. Horn FK, Tornow RP, Juenemann AG, Laemmer R, Kremers J.Perimetric measurements with flicker defined form stimulation in comparison to conventional perimetry and retinal nerve fiber measurements.Invest Ophthalmol Vis Sci. 2013.
8. Kiddee W, Tantisarasart T, Wangsupadilok B. Performance of optical coherence tomography for distinguishing between normal eyes, glaucoma suspect and glaucomatous eyes. J Med Assoc Thai. 2013;96(6):689-95.
9. Zeried FM, Osuagwu UL.Changes in retinal nerve fiber layer and optic disc algorithms by optical coherence tomography in glaucomatous Arab subjects. Clin Ophthalmol. 2013;7:1941-9.
10. Hwang YH, Kim YY, Kim HK, Sohn YH. Ability of cirrus high-definition spectral-domain optical coherence tomography clock-hour, deviation, and thickness maps in detecting photographic retinal nerve fiber layer abnormalities. Ophthalmology. 2013;120(7):1380-7.
11. Suh MH, Kim SK, Park KH, Kim DM, Kim SH, Kim HC. Combination of optic disc rim area and retinal nerve fiber layer thickness for early glaucoma detection by using spectral domain OCT. Graefes Arch ClinExpOphthalmol. 2013;251(11):2617-25.
12. Chung HJ, Park CK. Factors Determining the Peripapillary Retinal Nerve Fiber Distribution. J Glaucoma; 2013.
13. Zhao JJ, Zhuang WJ, Yang XQ, Li SS, Xiang W. Peripapillary retinal nerve fiber layer thickness distribution in Chinese with myopia measured by 3D-optical coherence tomography. Int J Ophthalmol. 2013;6(5):626-31.
14. Roh KH, Jeoung JW, Park KH, Yoo BW, Kim DM. Long-term reproducibility of cirrus HD optical coherence tomography deviation map in clinically stable glaucomatous eyes.Ophthalmology. 2013;120(5):969-77.
15. Iverson SM, Sehi M. The comparison of manual vs automated disc margin delineation using spectral-domain optical coherence tomography. Eye (Lond). 2013;27(10): 1180-7.
16. Chang RT, Knight OJ, Feuer WJ, Budenz DL. Sensitivity and specificity of time-domain versus spectral-domain optical coherence tomography in diagnosing early to moderate glaucoma. Ophthalmology. 2009;116(12):2294-9.

17. Mayama C, Saito H, Hirasawa H, Konno S, Tomidokoro A, Araie M, Iwase A, Ohkubo S, Sugiyama K, Otani T, Kishi S, Matsushita K, Maeda N, Hangai M, Yoshimura N. Circle- and grid-wise analyses of peripapillary nerve fiber layers by spectral domain optical coherence tomography in early-stage glaucoma.Invest Ophthalmol Vis Sci. 2013;54(7):4519-26.
18. Liu S, Wang B, Yin B, Milner TE, Markey MK, McKinnon SJ, Rylander HG 3rd. Retinal nerve fiber layer reflectance for early glaucoma diagnosis. J Glaucoma. 2014;23(1):e45-52.
19. Reznicek L, Seidensticker F, Mann T, Hübert I, Buerger A, Haritoglou C, Neubauer AS, Kampik A, Hirneiss C, Kernt M. Correlation between peripapillary retinal nerve fiber layer thickness and fundus autofluorescence in primary open-angle glaucoma. ClinOphthalmol. 2013;7:1883-8.

CHAPTER

6

Progression Analysis

■ INTRODUCTION

The optical coherence tomography (OCT) has been proved to be of immense benefit in determining progression of glaucoma as opposed to ascertaining a diagnosis.[1-5] The ability of the OCT to pick up structural changes at micron levels possibly outdoes that of the human eye to spot changes in serial fundus photographs and more importantly tends to negate to a large extent inter-observer variability with regard to fundus examinations. The fact that structural change tends to preceed functional change also means that the OCT is able to help diagnose and follow-up preperimetric glaucoma. In this respect, progression analysis is an extremely important module of the OCT with regard to glaucoma. The software module used is called guided progression analysis (GPA). It primarily evaluates the retinal nerve fiber layer (RNFL) changes over time and some optic nerve head parameters. However, serial ganglion cell layer thinning and the baseline values can also predict progression.[6-8]

In the presence of severe RNFL loss, it has been shown that the GPA program plays an important role in detecting progression.[9] Even in a scenario to the contrary, in the presence of mild visual field defects, it is the OCT RNFL which is able to detect progression better than the fields.[10] Loss of the inferior RNFL and abnormal optic nerve head topography are other OCT parameters associated with glaucoma progression.[11]

■ TECHNOLOGY

Guided progression analysis (GPA) compares RNFL thickness measurements from optic disc cube scans (3 to 8 scans) over time and evaluates for statistically significant changes. It requires a minimum of three scans since two are taken to be the baseline against which the subsequent scans are compared. The analysis displays RNFL thickness maps in chronological order and derives RNFL thickness change maps, average RNFL thickness graphs representing rate of change, and RNFL thickness profiles comparing the current scan to the baseline (Figure 6.1).

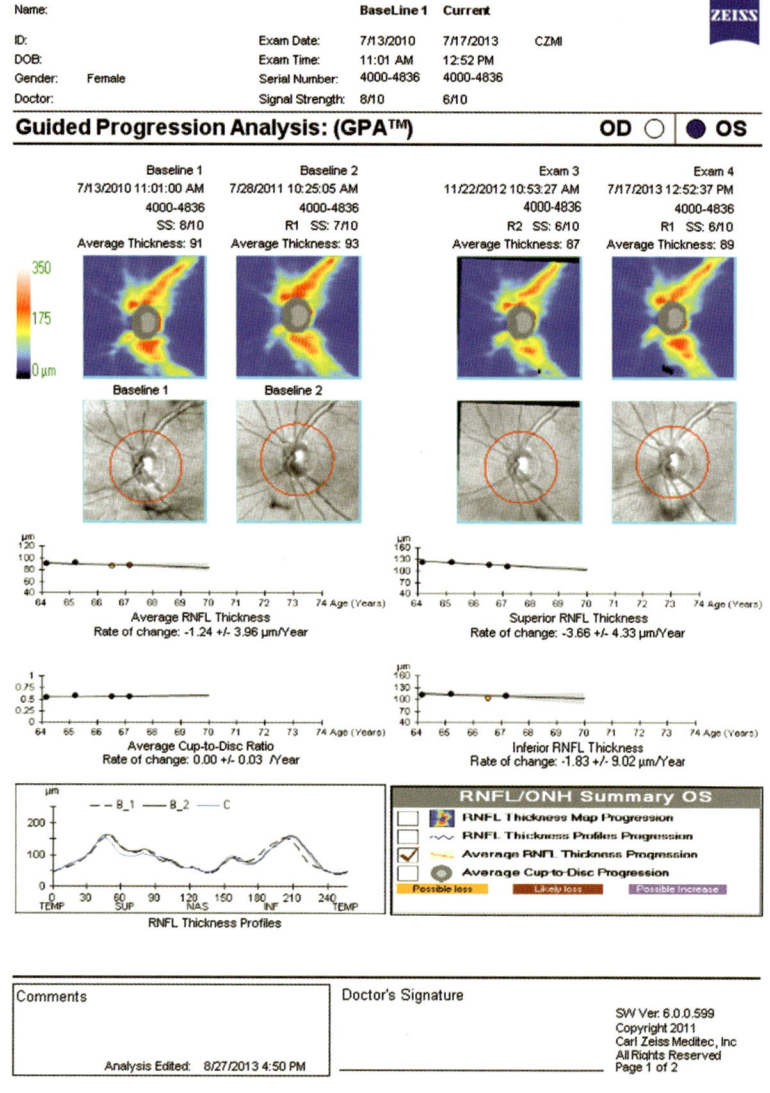

Fig. 6.1 The change analysis part of a typical GPA printout. Note the various graphs produced

Statistically significant changes are flagged for possible or likely RNFL loss or gain.

For the purposes of analysis, GPA automatically selects the appropriate scans having highest signal strength from the most recent 8 visits. Scans with signal strength of 5 or lower are excluded by default though they may be manually included.

The two baselines are either included in the linear regression to determine rate of change, confidence limits on that rate, and statistical significance of the trend for the summary parameters or they may be used for evaluating *possible* or *likely* change.

Progression Analysis

In the latest version of the OCT, the RNFL change analysis is supplemented with an ONH change analysis.

■ METHOD

The first two appropriate scans (having signal strength >5) establish the baseline, though manual selection of any of the scans may be done (Figure 6.2). The software labels all scans as either B for baseline or a green checkmark for acceptable scans

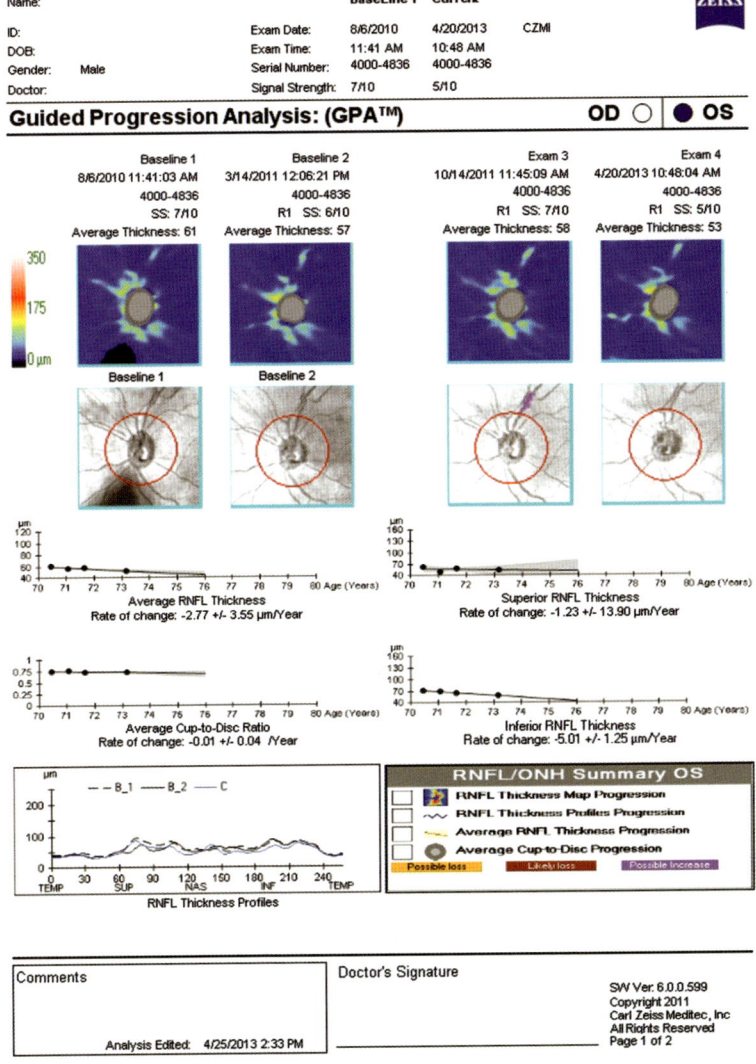

Fig. 6.2 This figure demonstrates how two baseline OCT results are used to perform a guided progression analysis. Note the two baseline scans marked as baseline 1 and baseline 2 along with the dates of the scans

or a red cross for poor quality scans which are not suitable for GPA. An operator may manually check or cross out any scan and then select the ones required ones for the progression analysis.

■ INTERPRETATION

The GPA printout contains the RNFL.
- *RNFL thickness map:* Topographic representation of each exam. Interpreted in a manner similar to an RNFL printout, i.e warm colors are thicker and cool colors represent thinner areas (Figure 6.3).
- *RNFL thickness change map:* This highlights the change in RNFL thickness over time. If the change has happened in one exam, it is marked as yellow, and if it is persistent in many exams, it is marked in red (Figure 6.4).
- *Average RNFL thickness graphs:* The software plots the RNFL thickness over the various exams into a graph. This could be average RNFL thickness or superior RNFL thickness or inferior RNFL thickness. If there is a change noted, it is marked as yellow if there is a change in one exam and red if the change is seen in multiple exams. The graph also mentions the rate of change and if significant, it provides the p value (Figure 6.5).
- *Average cup-to-disc graph:* This graph depicts the change in cup disc ratio over time and provides the rate of change of cupping. In a similar manner to RNFL graphs, it labels possible progression as yellow and likely progression as red. This is actually is supplement to the RNFL progression since it is derived from the ONH scans (Figure 6.6).

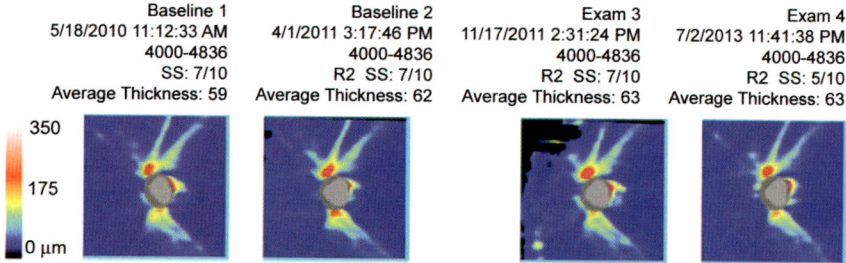

Fig. 6.3 Retinal nerve fiber layer thickness map

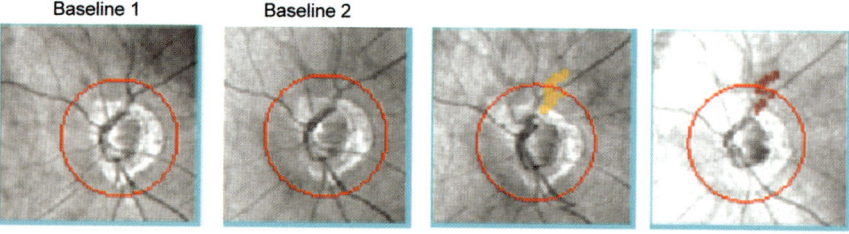

Fig. 6.4 RNFL thickness change map. Note the area marked in yellow in the third picture which is indicating a change in one exam. When this change persists in the subsequent exam, note how the marking changes to red

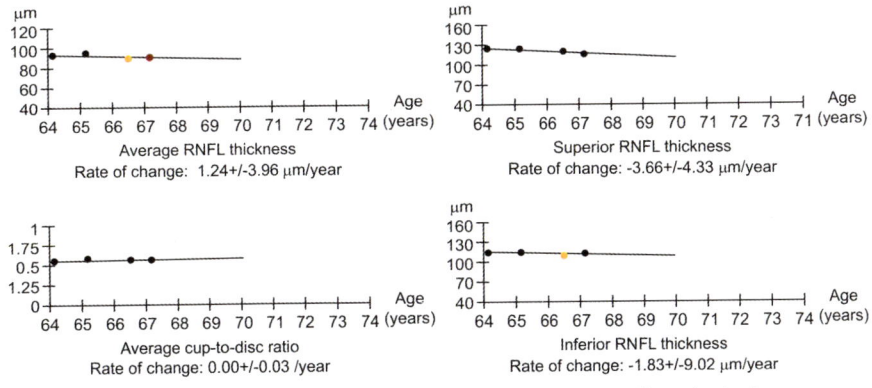

Fig. 6.5 RNFL thickness progression graphs. Note the yellow dot in the inferior RNFL thickness graph

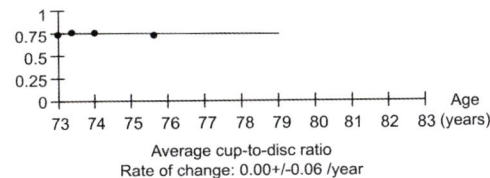

Fig. 6.6 Average Cup-to-disc ratio progression graph

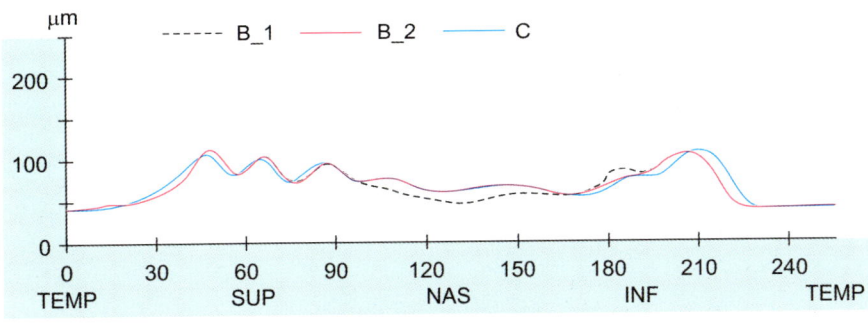

Fig. 6.7 RNFL profile (TSNIT) thickness progression graph

- *RNFL profile graph:* Sequential TSNIT profile is plotted for RNFL on a graph and it is labeled as red for likely RNFL loss and yellow for possible RNFL loss. The TSNIT profile of both baselines is also plotted (Figure 6.7).
- *RNFL summary:* This summarizes the GPA report. The RNFL thickness map progression is an indicator of focal change and a tick mark in this box depicts focal RNFL loss. The RNFL thickness profiles progression is best for broader focal change and if there is any, this box is checked. The average RNFL thickness progression is best for diffuse change and a checkmark next to this indicates a diffuse loss of RNFL (Figure 6.8).

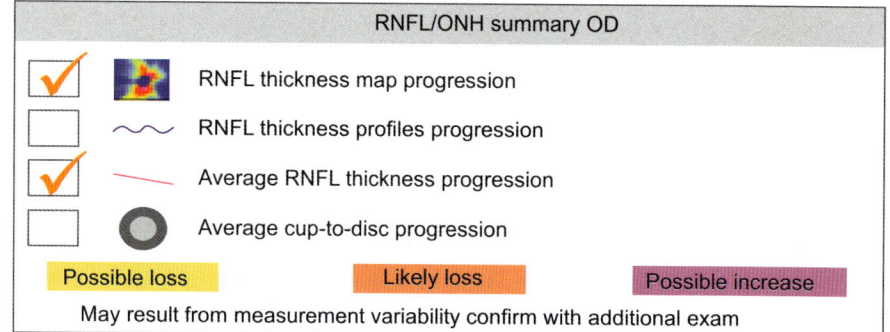

Fig. 6.8 RNFL/ONH summary showing RNFL loss in both a focal and a diffuse manner. There is likelihood of glaucoma progression. The yellow colored checkmarks means there is possible loss recorded in one follow-up visit

Key Points
- Progressive structural changes clinch the diagnosis of glaucoma.
- Structural change often precedes functional change.
- OCT can be used to evaluate progression of the optic nerve head and RNFL.
- It is important to maintain scan quality to interpret progression.

CASES

■ CASE 1: OCT VERSUS FUNDUS PHOTOGRAPHY

This is a patient with normal tension glaucoma with intraocular pressures always in the low teens. However there was suspected progression on the serial visual fields, particularly in the right eye, which was then confirmed by the guided progression analysis of the OCT. Notice how the GPA shows a gentle slope in the RNFL graph of both the eyes. The fundus however shows minimal change and may miss some of the progression. The patient was later found to have a carotid vasooccusive disease which explained the progression. This case exemplifies the correlation of structural and functional testing for diagnosing progression.

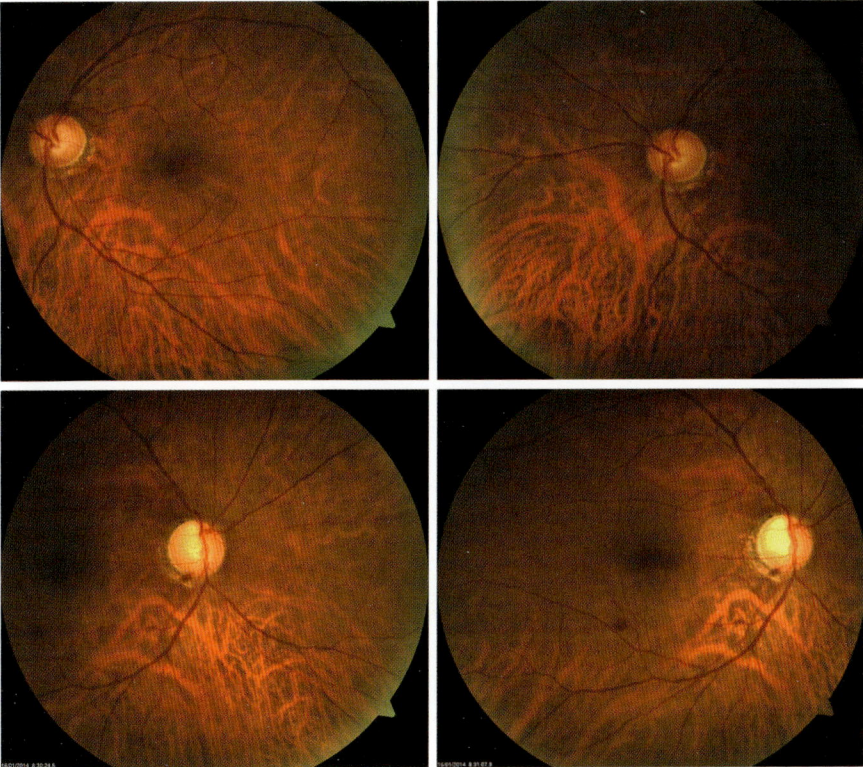

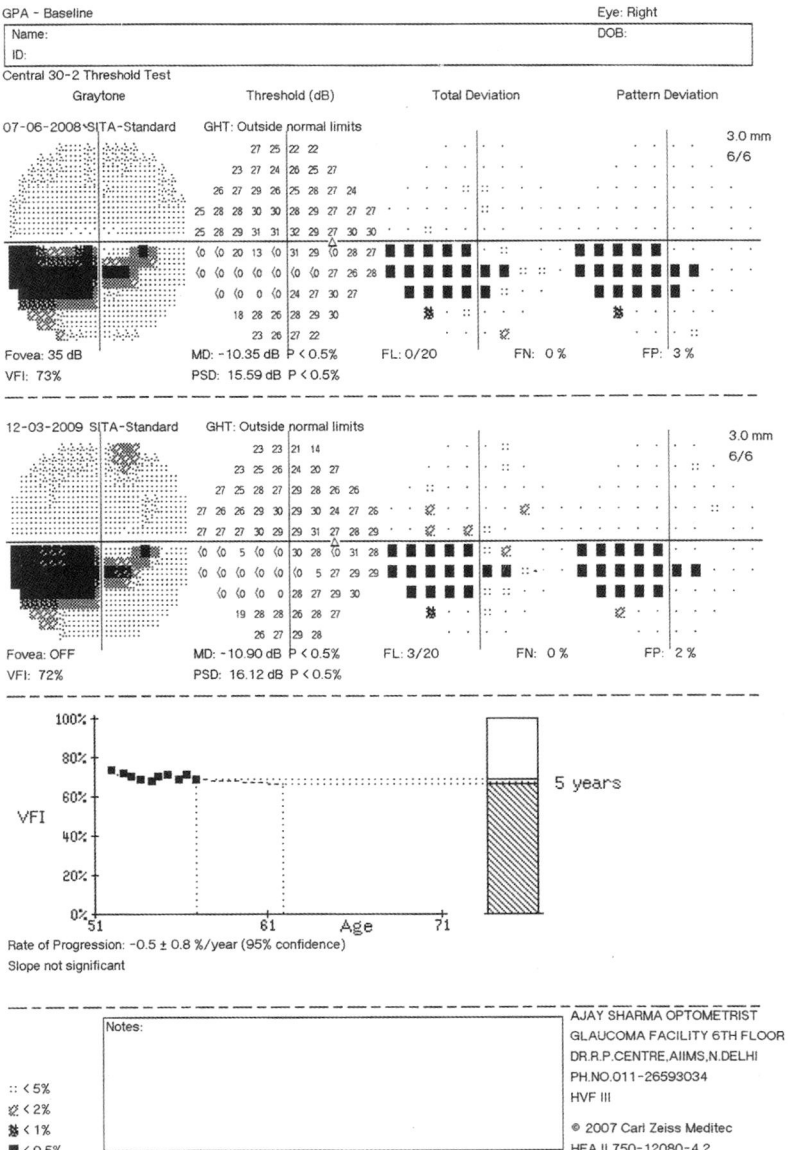

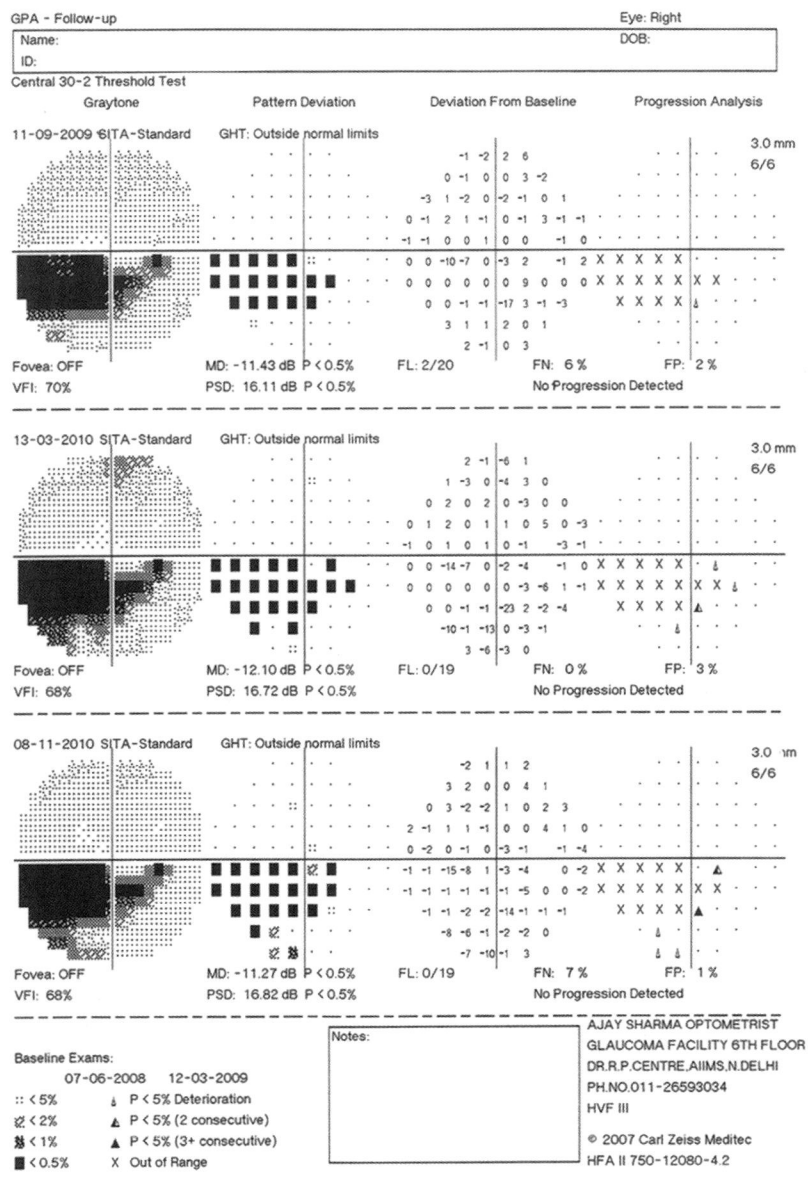

Optical Coherence Tomography in Current Glaucoma Practice

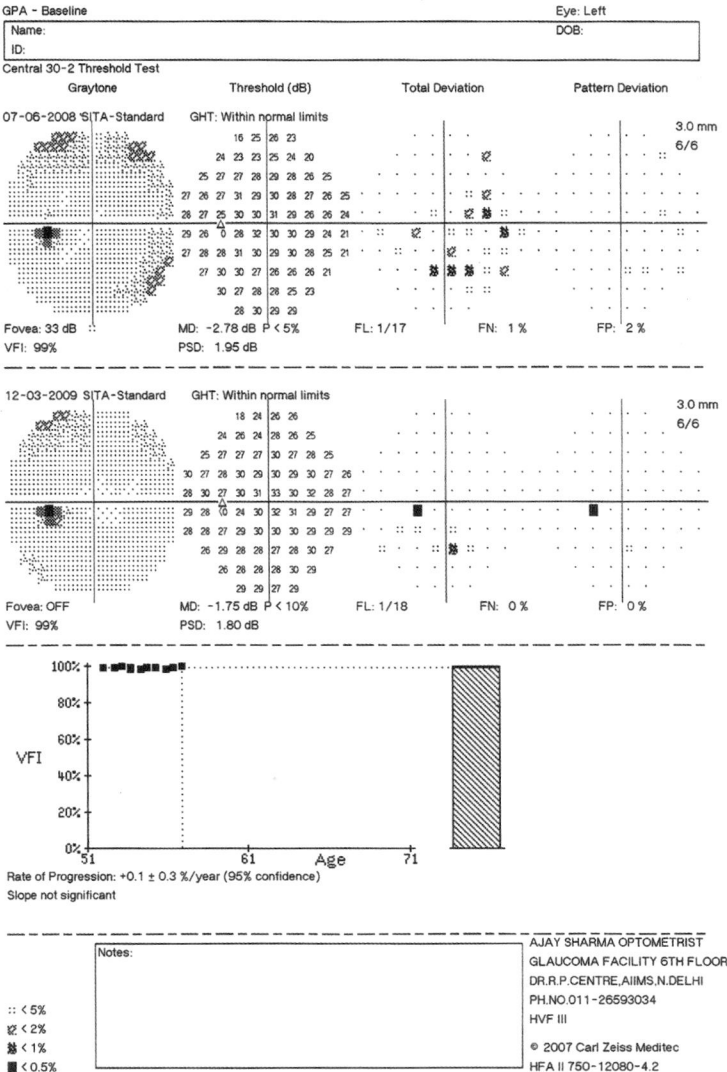

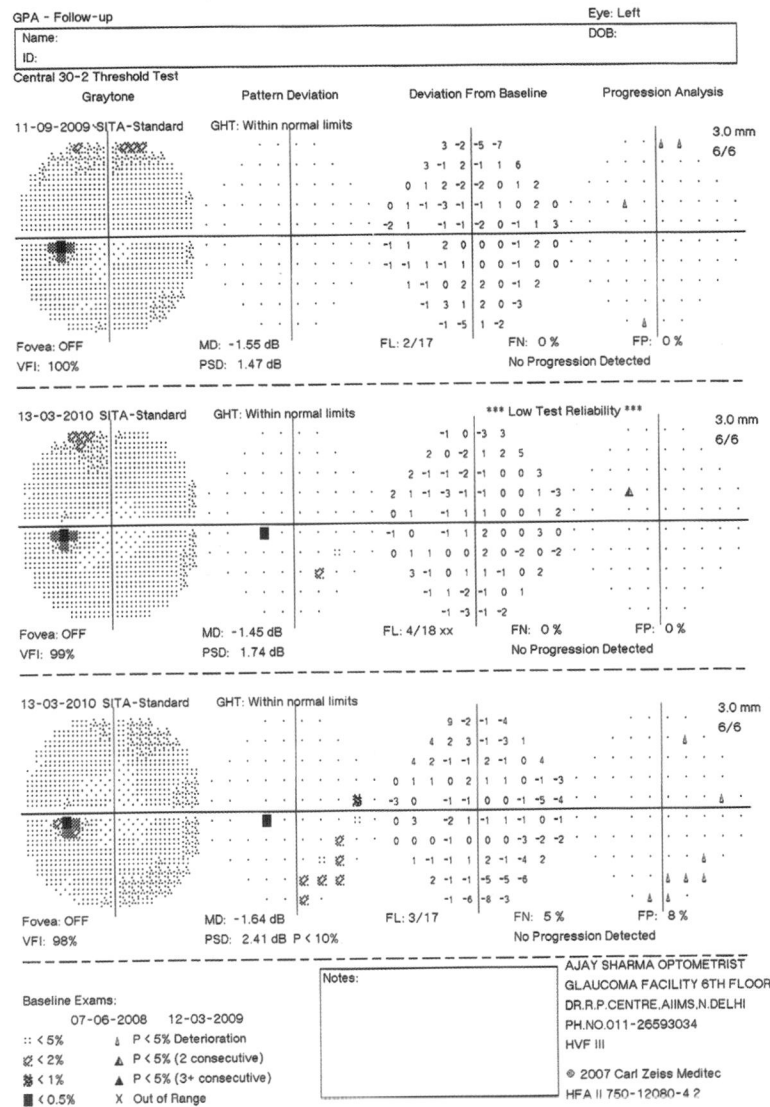

Name:
BaseLine 1 Current
ID: Exam Date: 3/13/2010 5/18/2013 CZMI
DOB: Exam Time: 10:07 AM 10:26 AM
Gender: Serial Number: 4000-4836 4000-4836
Doctor: Signal Strength: 7/10 6/10
ZEISS
Guided Progression Analysis: (GPA™)
OD OS
Baseline 1 3/13/2010 10:07:37 AM 4000-4836 SS: 7/10 Average Thickness: 75
Baseline 2 11/8/2010 10:49:19 AM 4000-4836 R2 SS: 8/10 Average Thickness: 79
Exam 3 11/17/2012 11:21:57 AM 4000-4836 R1 SS: 6/10 Average Thickness: 75
Exam 4 5/18/2013 10:26:09 AM 4000-4836 R2 SS: 6/10 Average Thickness: 72
350
175
0 µm
Baseline 1
Baseline 2
Age (Years)
Average RNFL Thickness
Rate of change: -1.27 +/- 4.01 µm/Year
Superior RNFL Thickness
Rate of change: -1.62 +/- 5.10 µm/Year
Average Cup-to-Disc Ratio
Rate of change: 0.01 +/- 0.06 /Year
Inferior RNFL Thickness
Rate of change: -1.30 +/- 1.43 µm/Year
B_1 B_2 C
TEMP SUP NAS INF TEMP
RNFL Thickness Profiles
RNFL/ONH Summary OS
RNFL Thickness Map Progression
RNFL Thickness Profiles Progression
Average RNFL Thickness Progression
Average Cup-to-Disc Progression
Possible loss
Likely loss
Possible increase
Comments
Analysis Edited: 5/18/2013 10:29 AM
Doctor's Signature
SW Ver. 6.0.0.599
Copyright 2011
Carl Zeiss Meditec, Inc
All Rights Reserved
Page 1 of 2

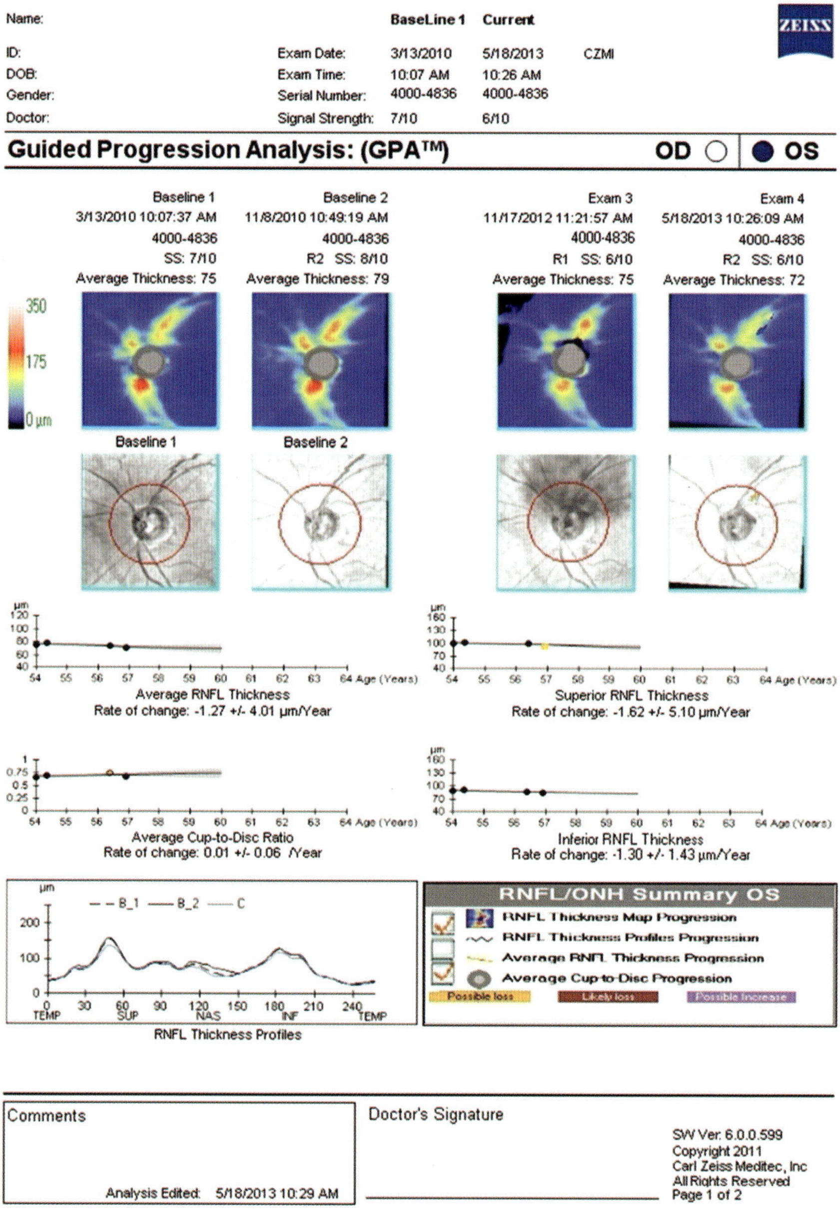

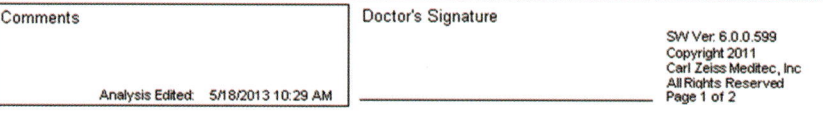

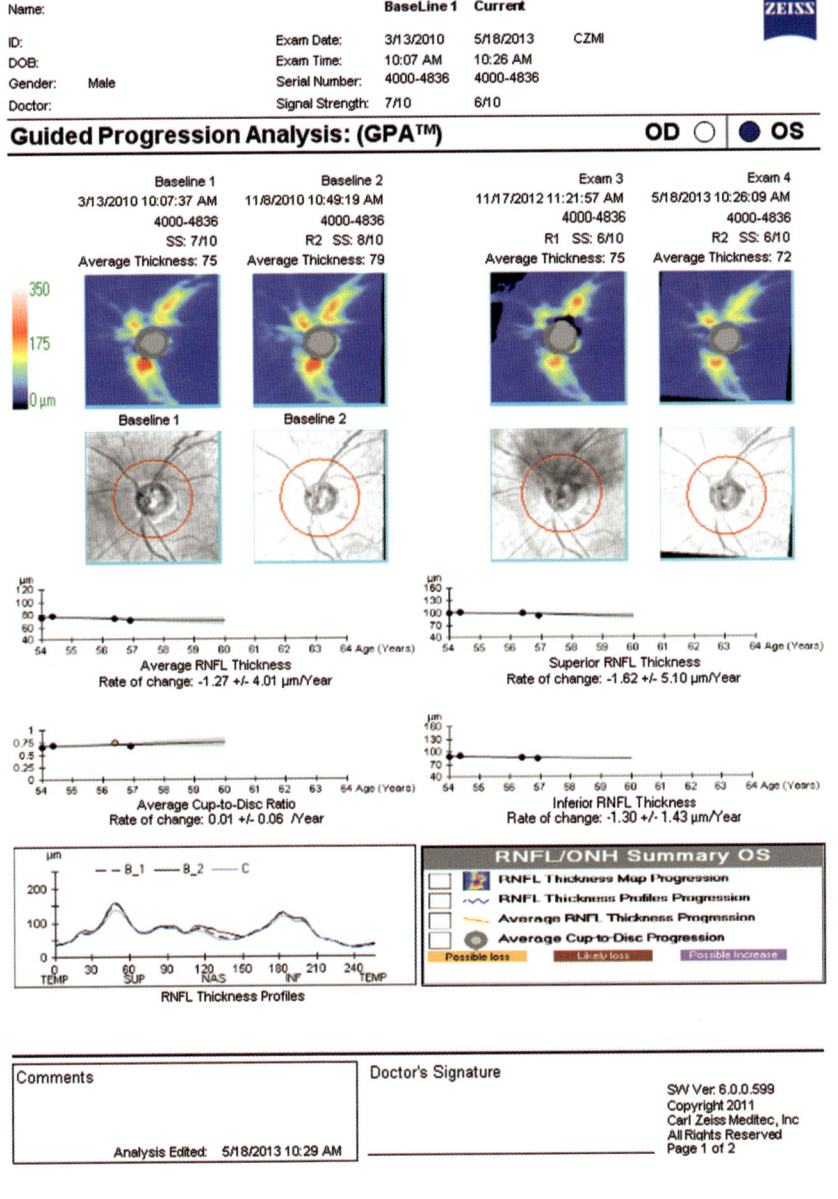

CASE 2: PSEUDOPROGRESSION

This is an elderly patient on regular follow-up for his angle closure glaucoma. Serial fields seem to show some progression in the fourth and fifth field of the left eye despite good intraocular pressure control. However, the OCT shows that the RNFL is stable in both the eyes and that the fields seem to show a pseudoprogression. The patient was noted to have a cataract and underwent cataract surgery and the sixth and seventh field of the left eye no longer showed progression. Thus, this case exemplifies that the OCT can help rule out pseudoprogression.

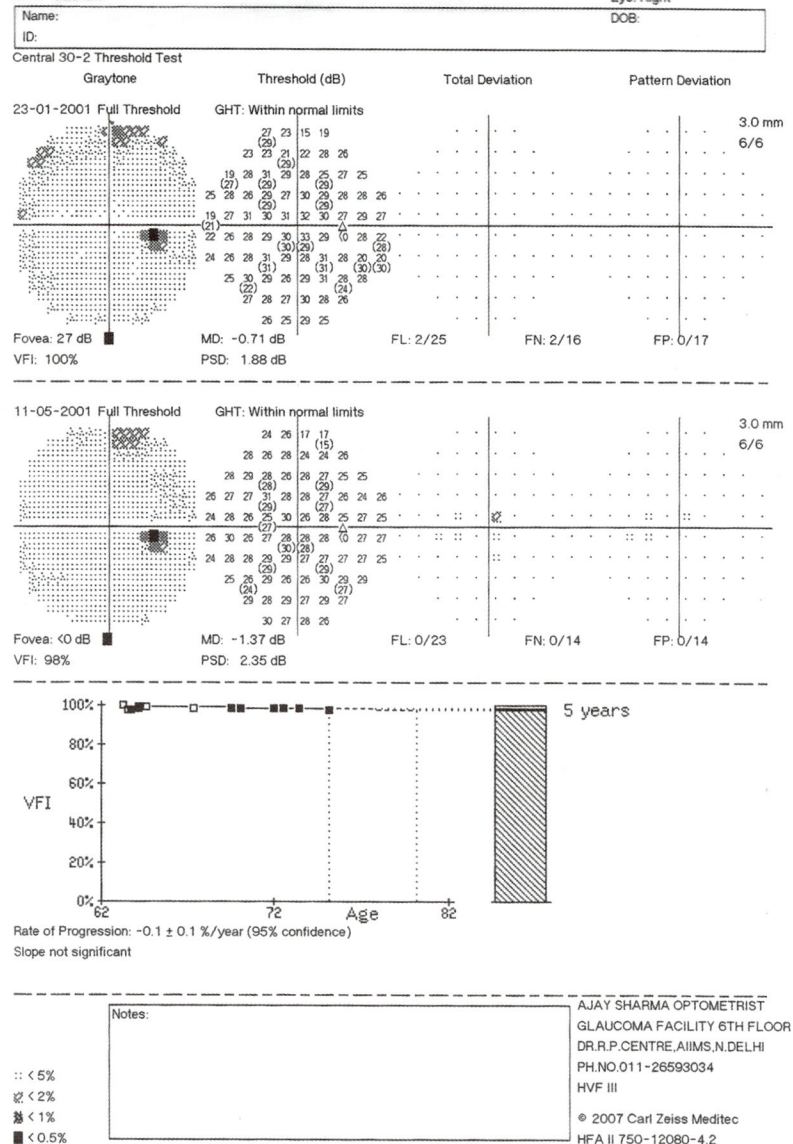

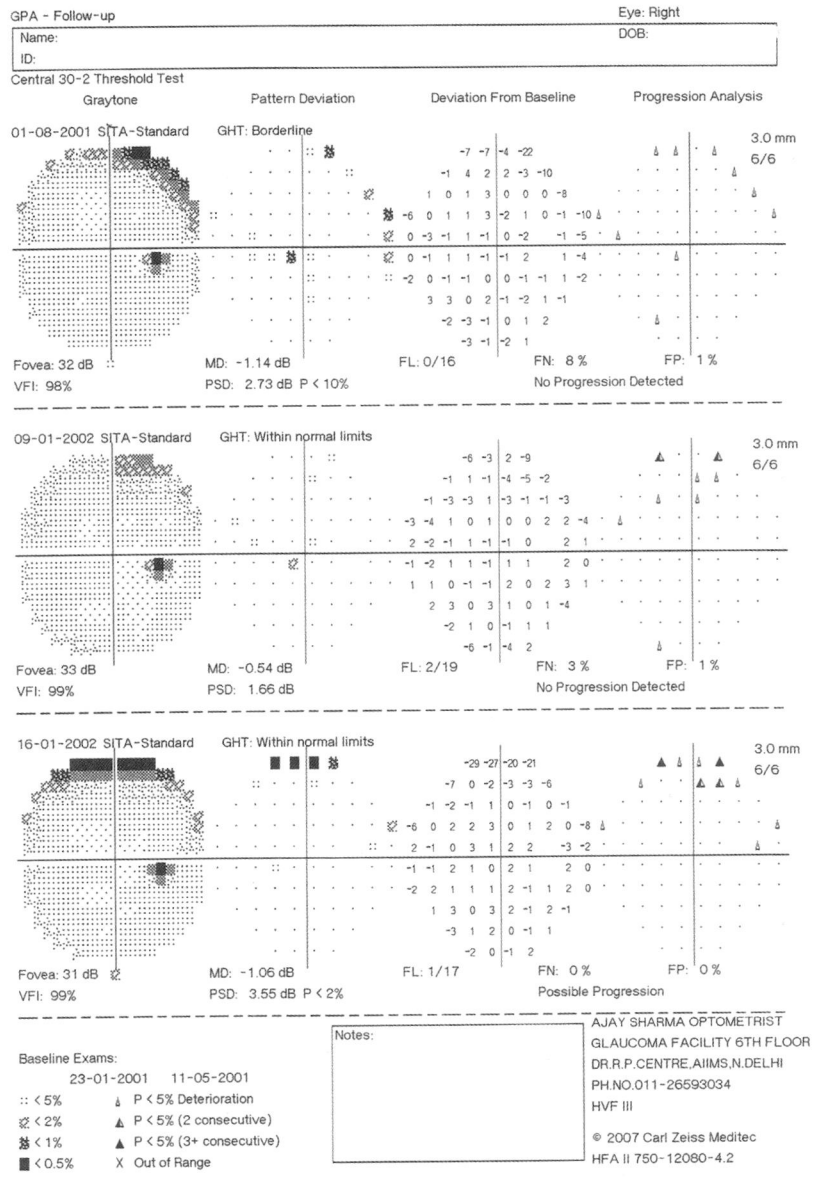

70 Optical Coherence Tomography in Current Glaucoma Practice

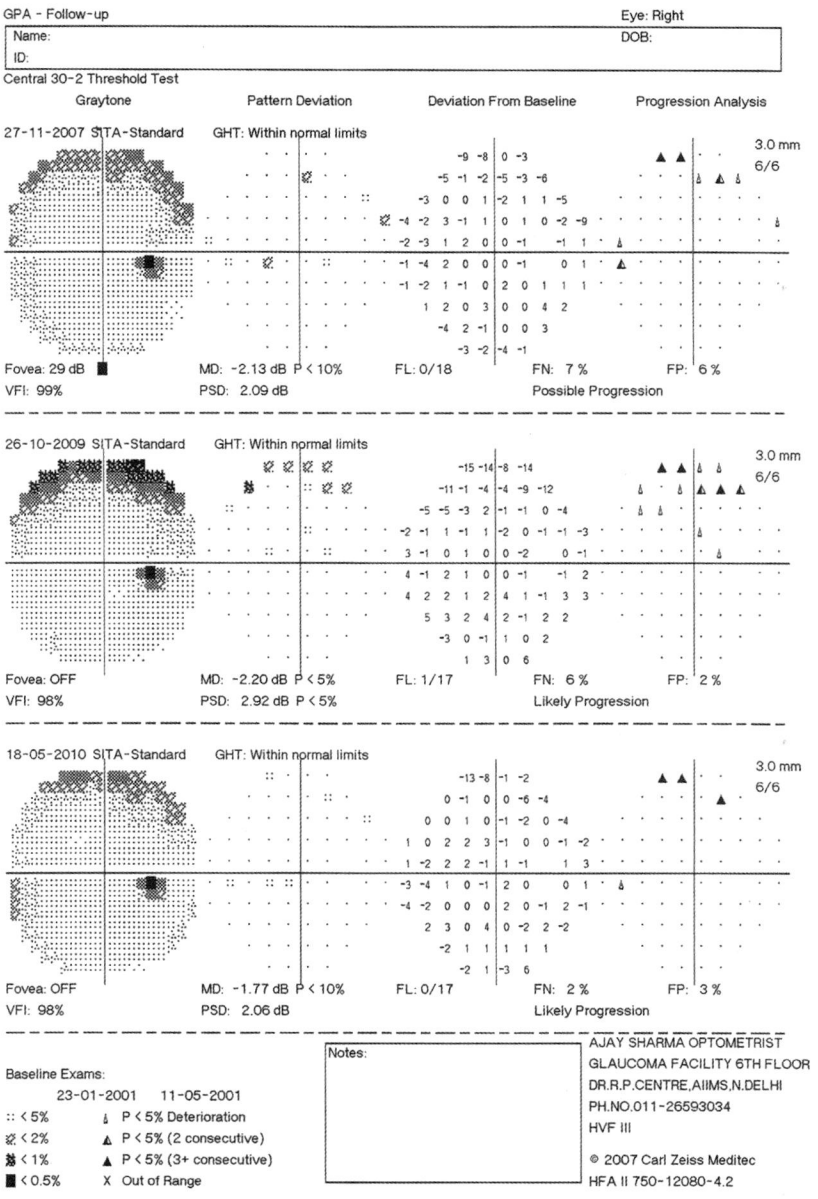

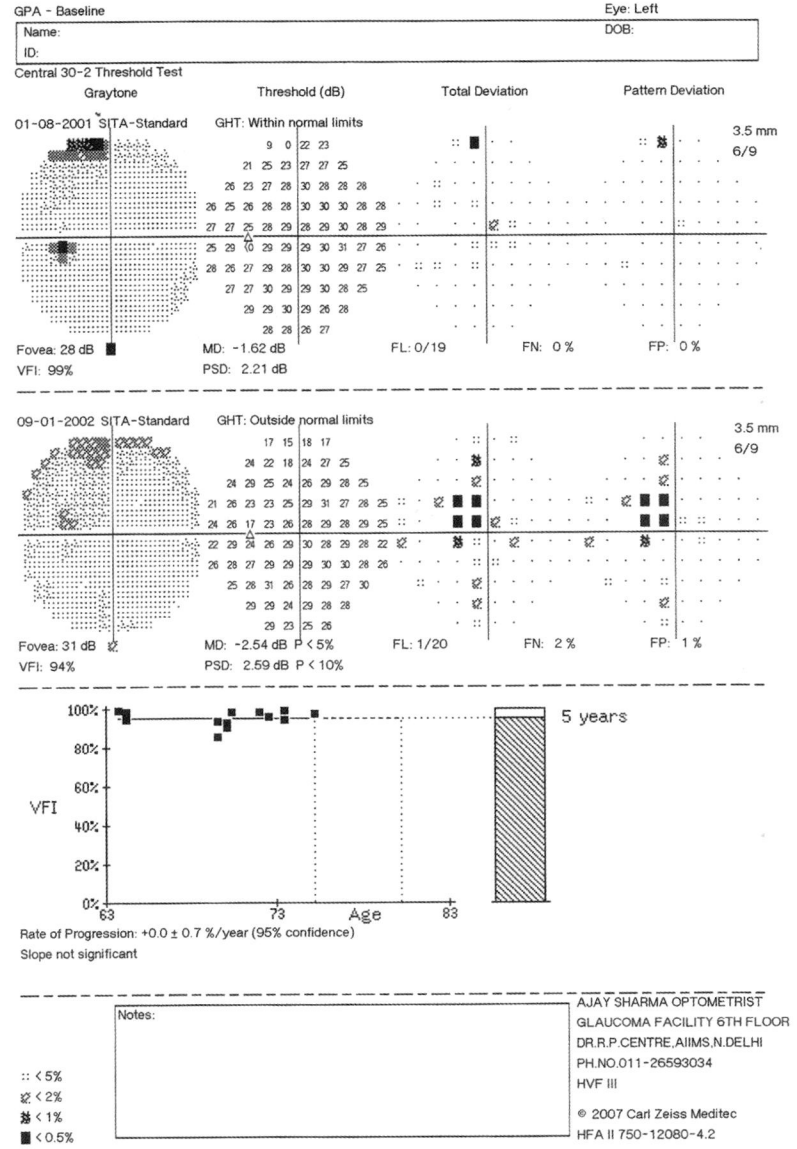

Optical Coherence Tomography in Current Glaucoma Practice

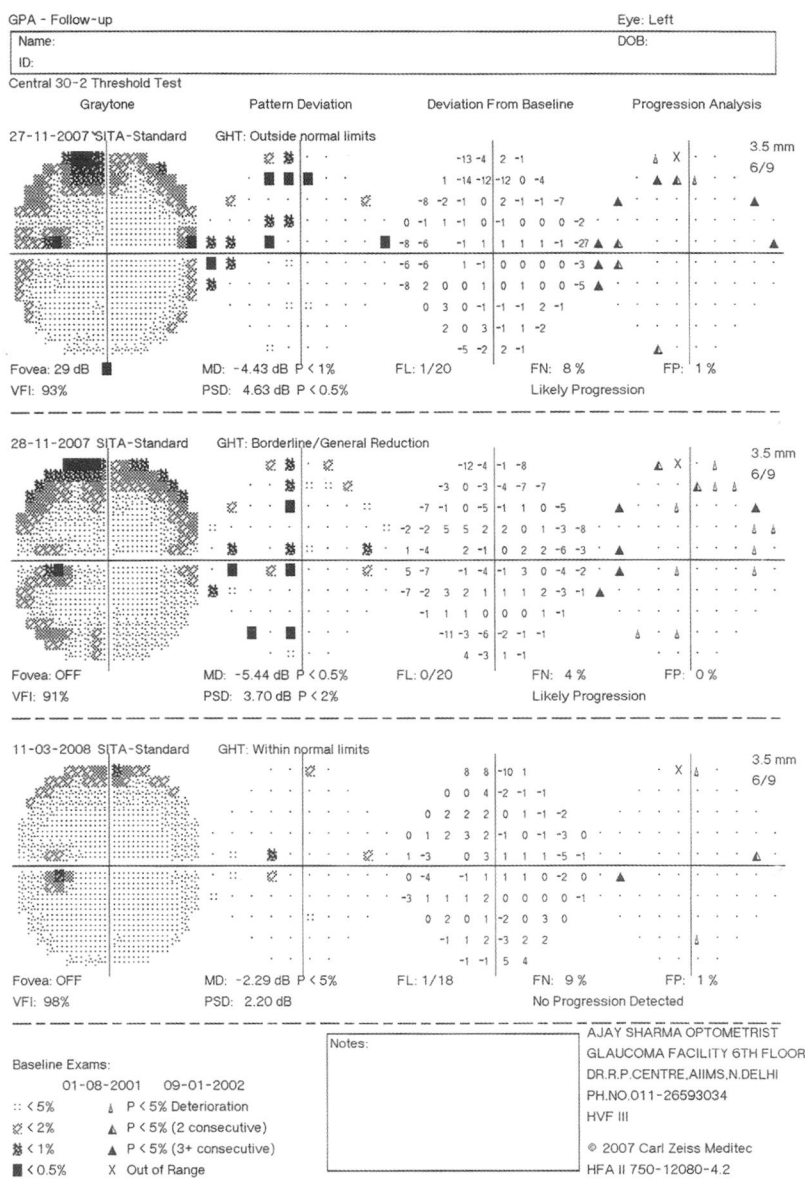

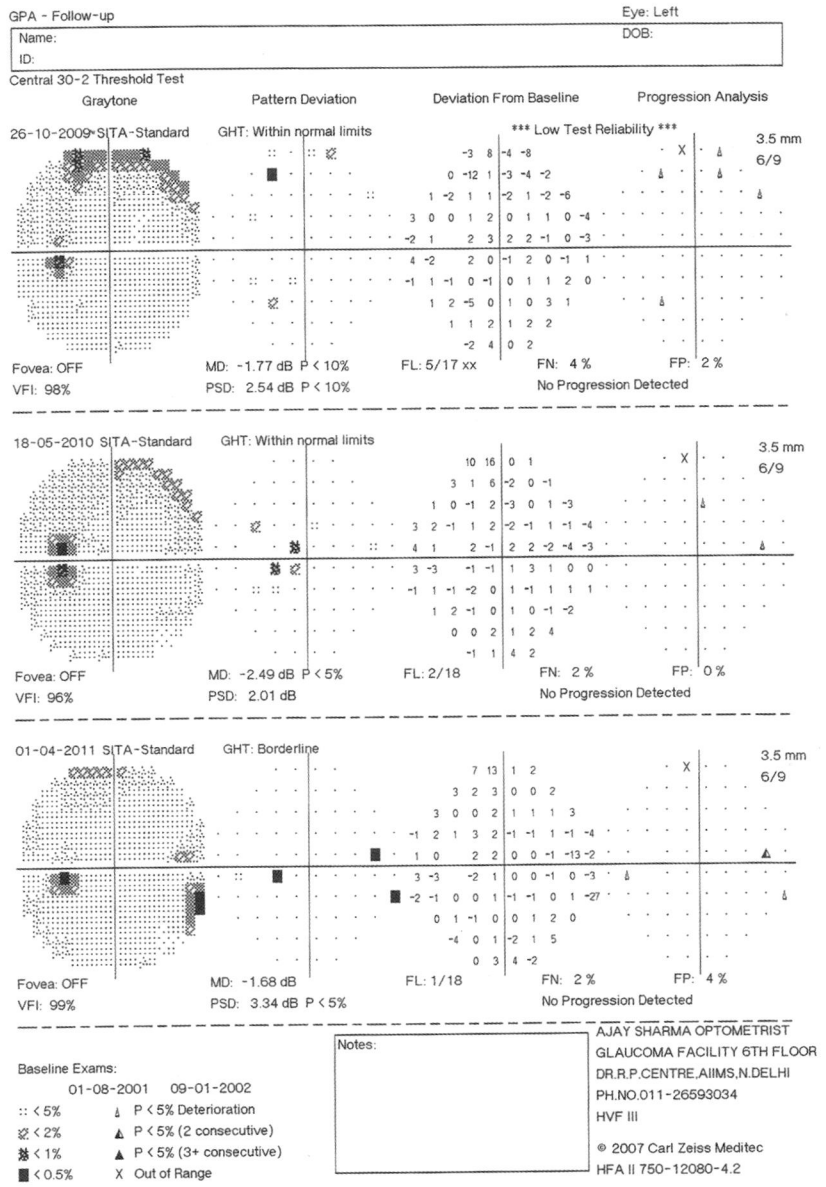

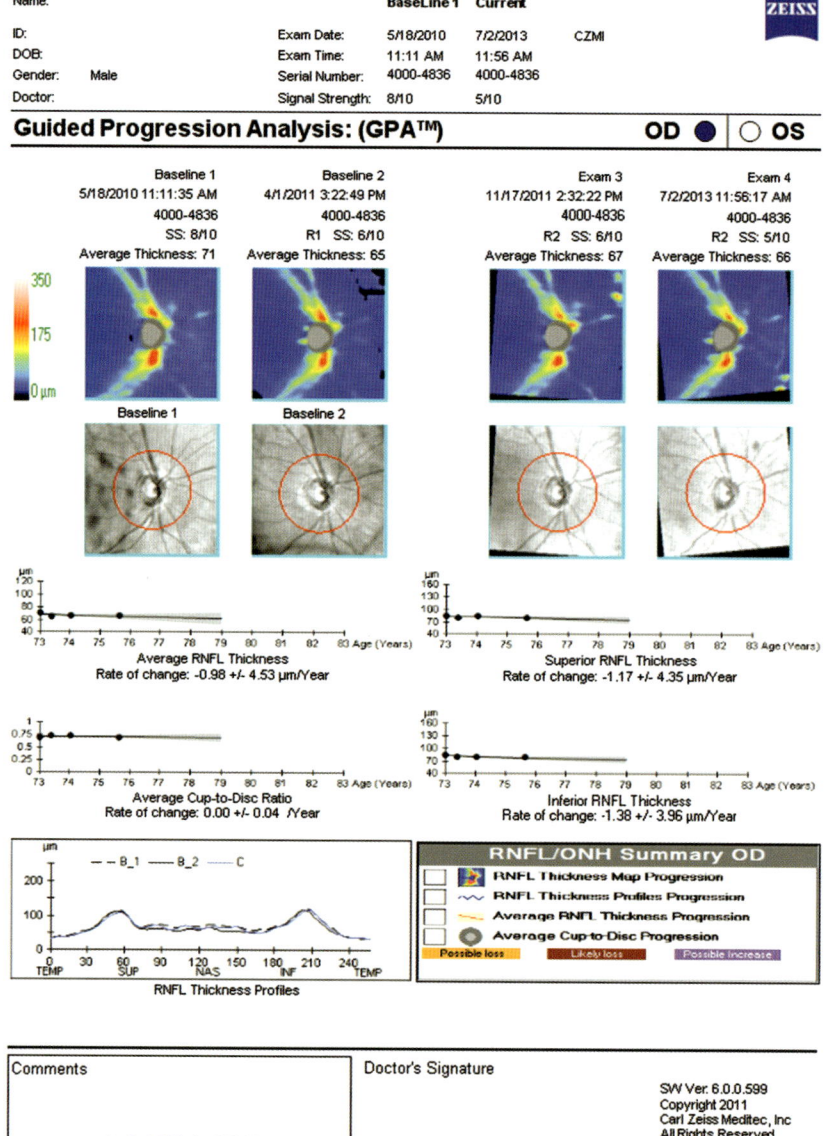

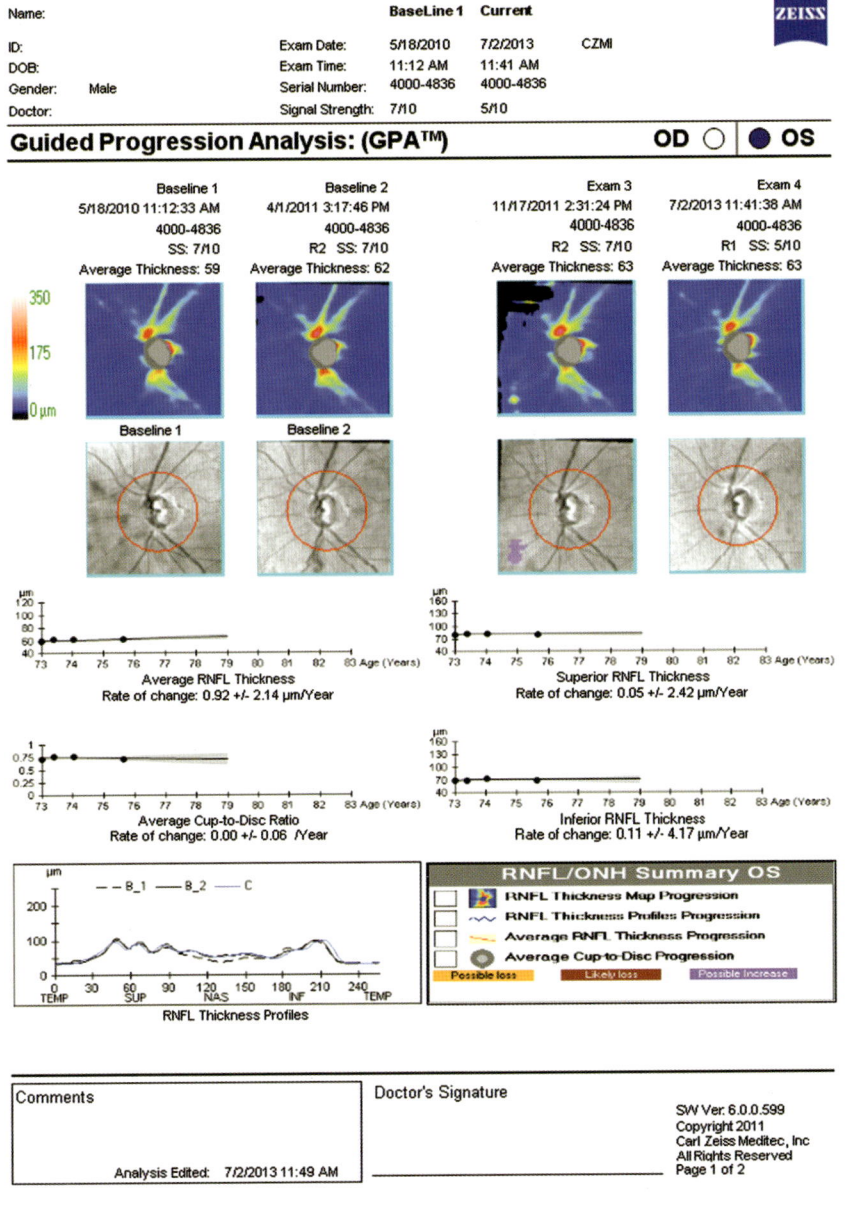

CASE 3: VARIABILITY ON OCT

This is a patient who shows some progression on visual fields though it is variable and inconsistent. In such a case where doubtful progression is observed on a functional test, structural tests such as the OCT should always be done to confirm or rule out disease progression. In this case, the OCT clearly rules out progression and infact there appears to be a pseudo improvement in RNFL. This may happen to variability in scanning or acquisition of a slightly different area each time or wrong baseline scan selection. In this case therefore, OCT indicates no progression but would have to be interpreted with caution.

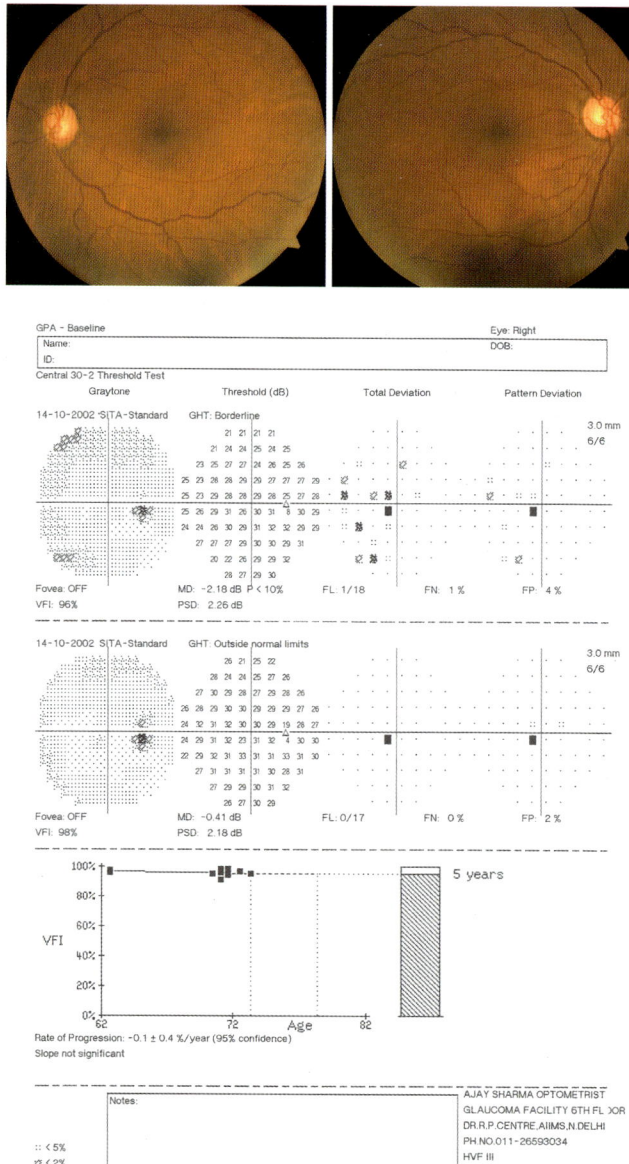

Progression Analysis

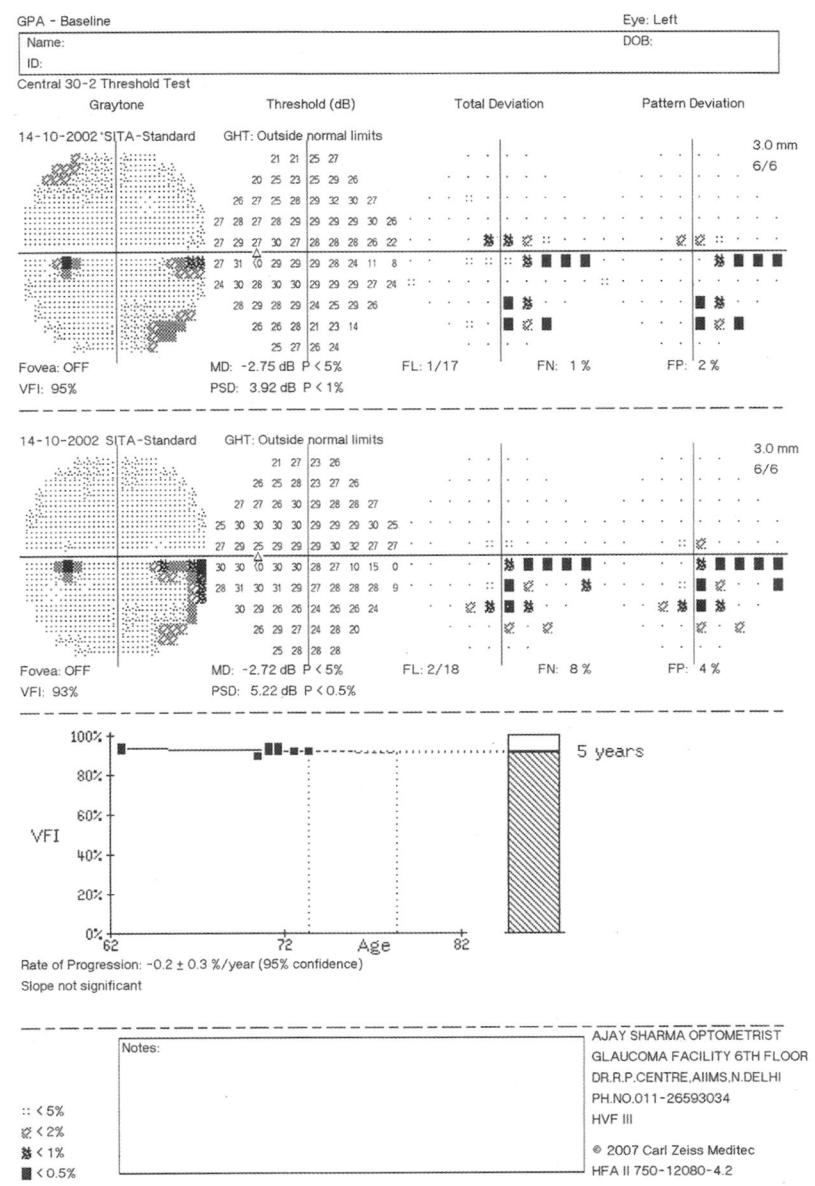

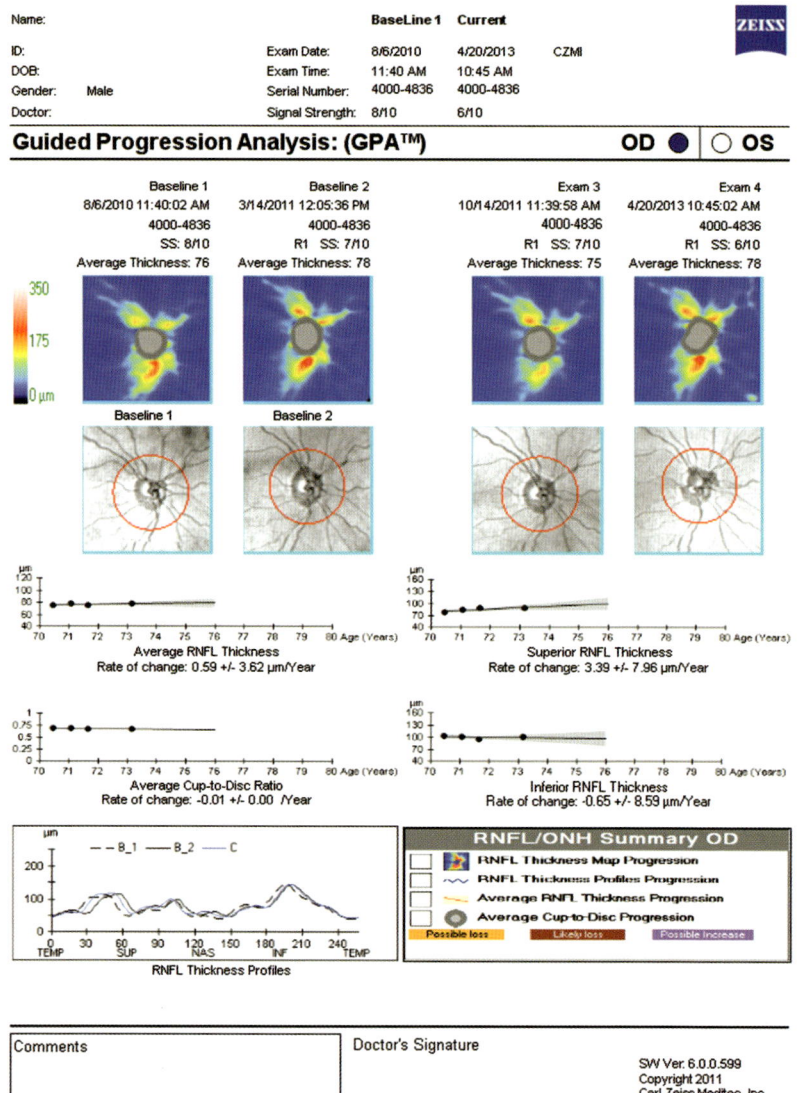

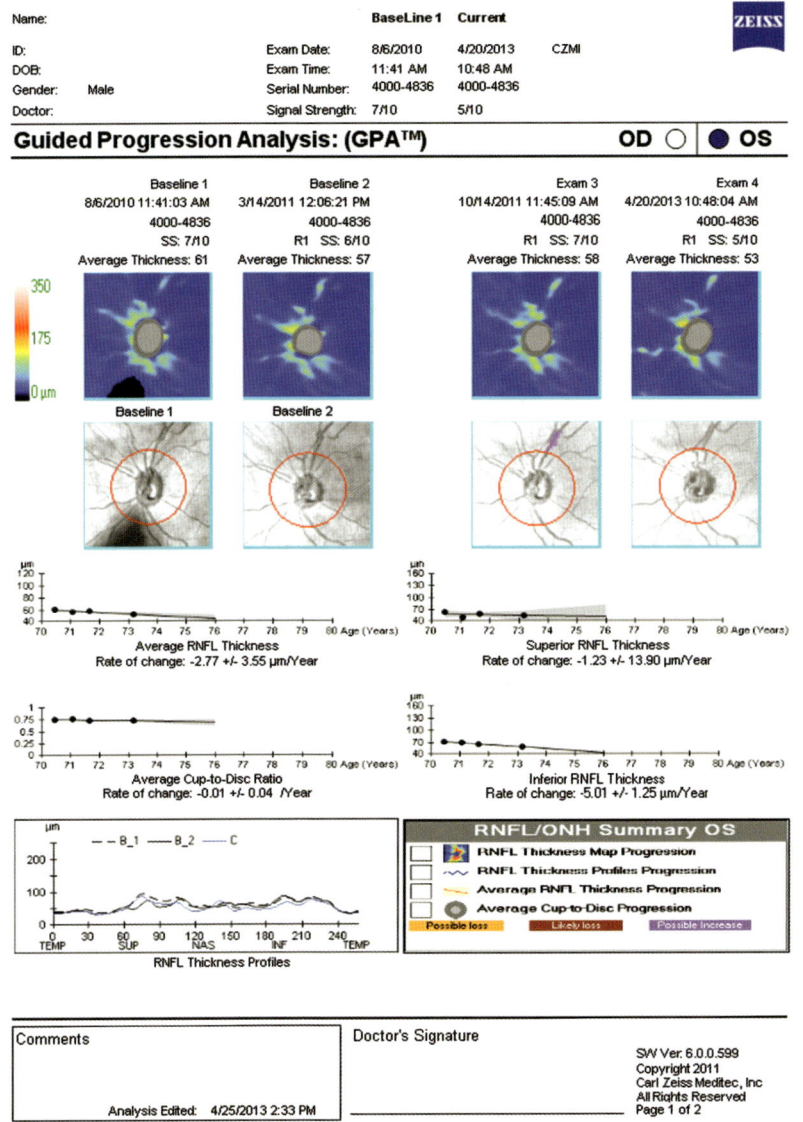

CASE 4: TRUE PROGRESSION: OCT CORRELATES WITH VISUAL FIELDS

This is a case of open angle glaucoma with progression on the visual fields of the left eye and relatively constant visual field loss of the right eye. The OCT shows progression in the left eye which corresponds with the progression on fields. However, in the right eye, the structural progression is seen to occur before the functional progression seen on the fields. This case exemplifies that the OCT is more sensitive to progression than the visual fields, particularly in moderate to advanced glaucoma and can help diagnose progression prior to actual field loss.

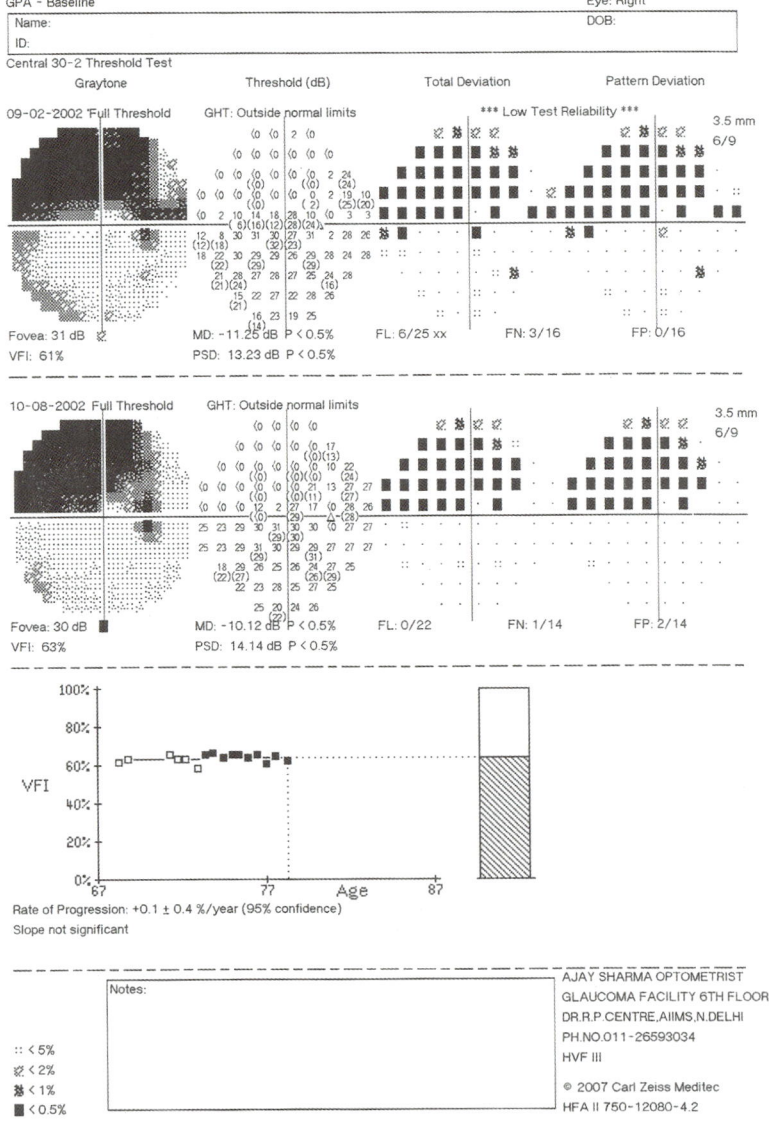

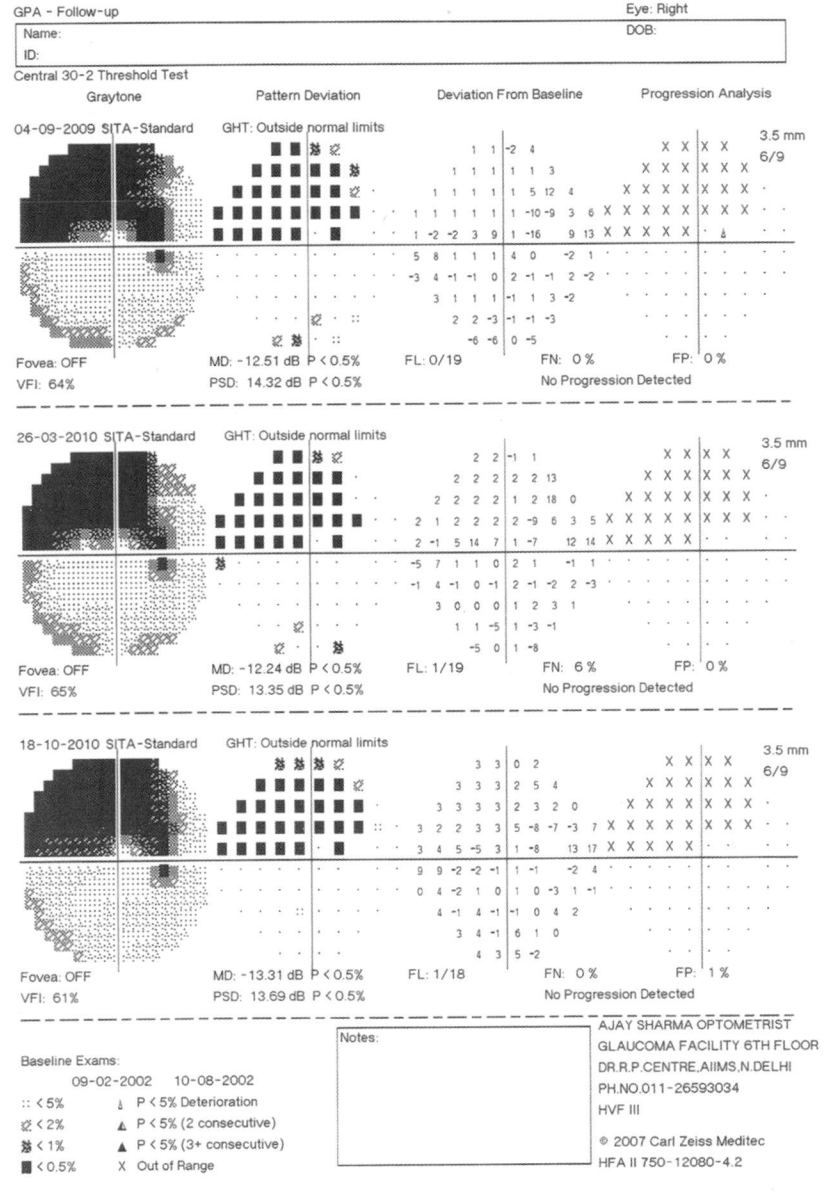

82 Optical Coherence Tomography in Current Glaucoma Practice

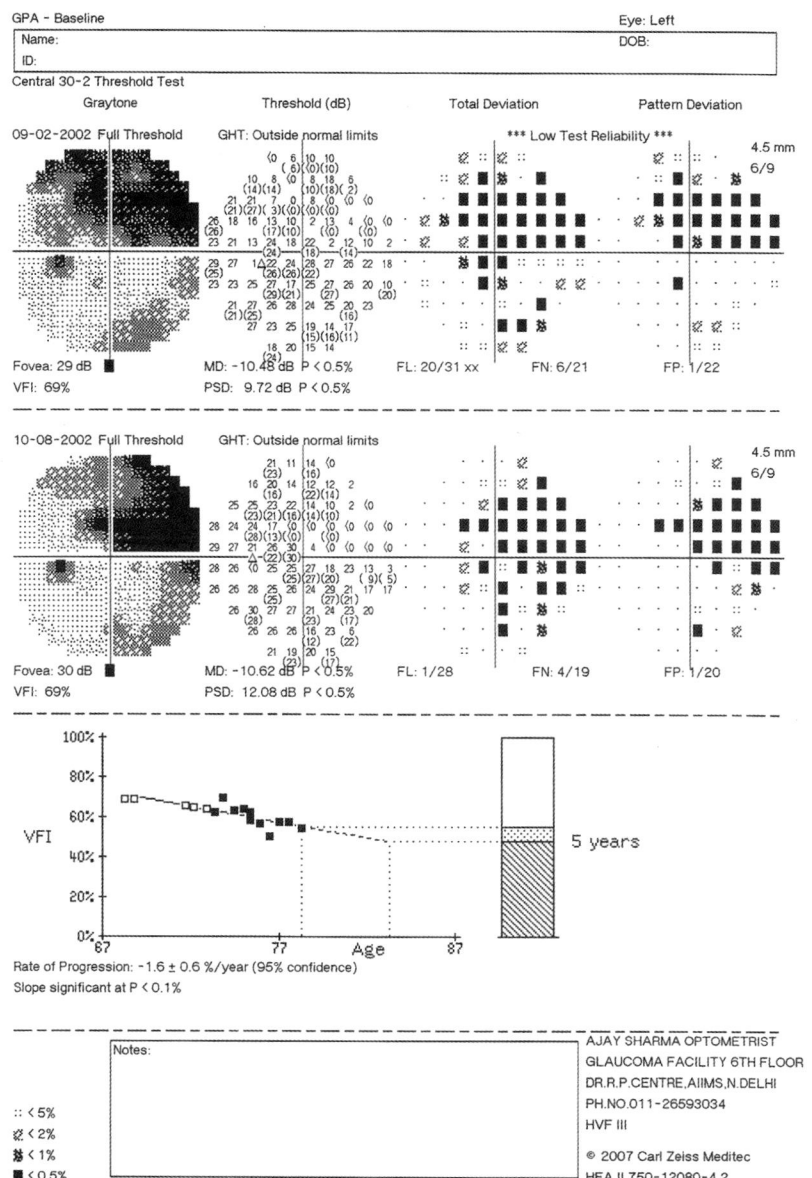

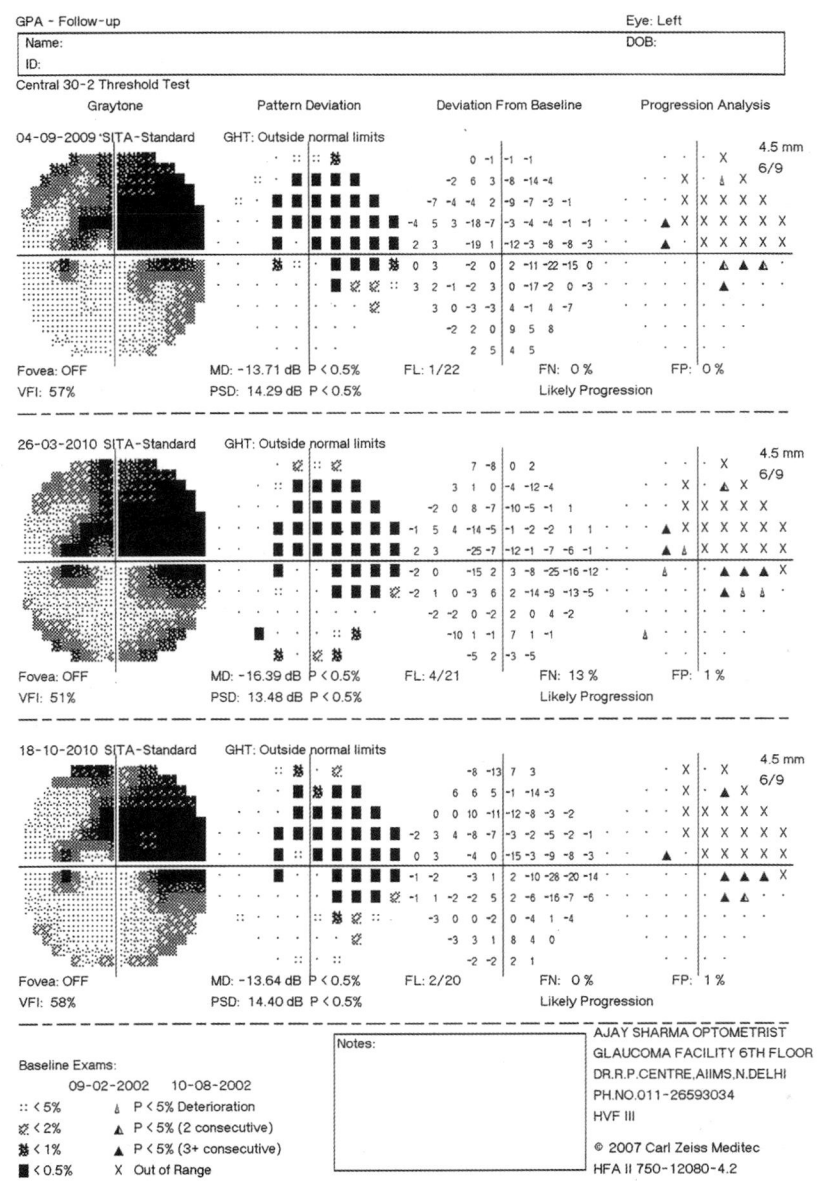

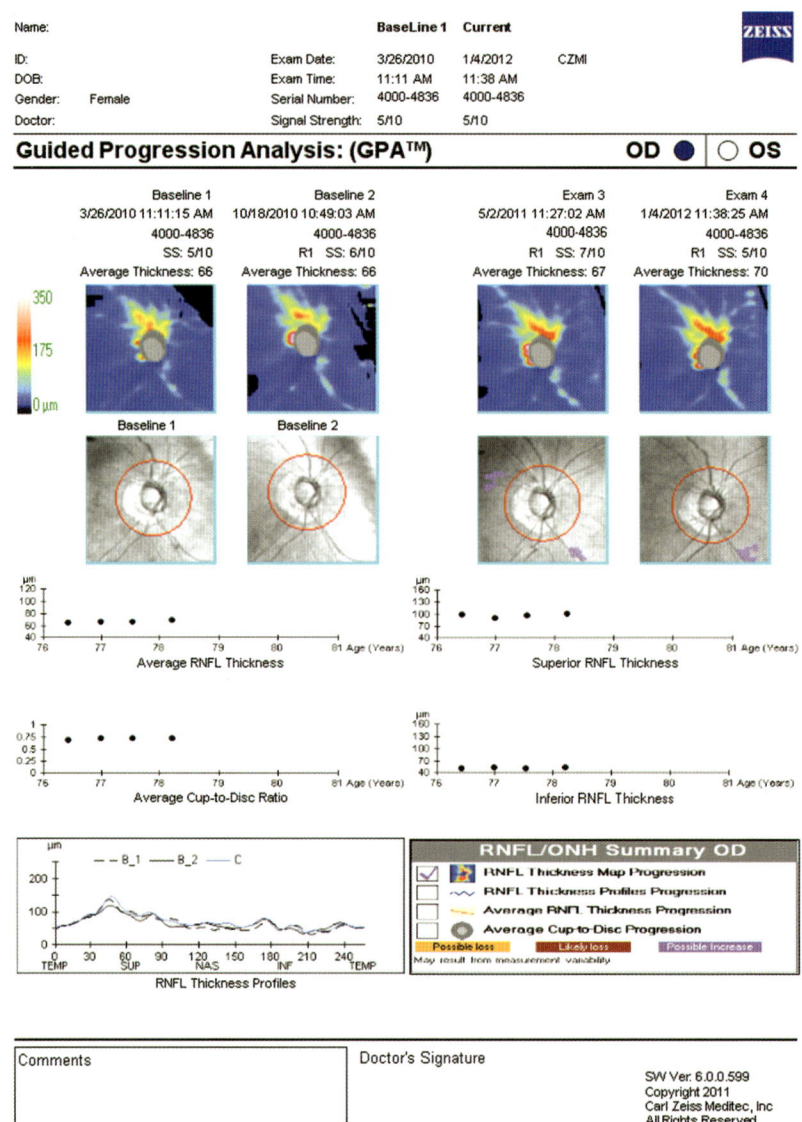

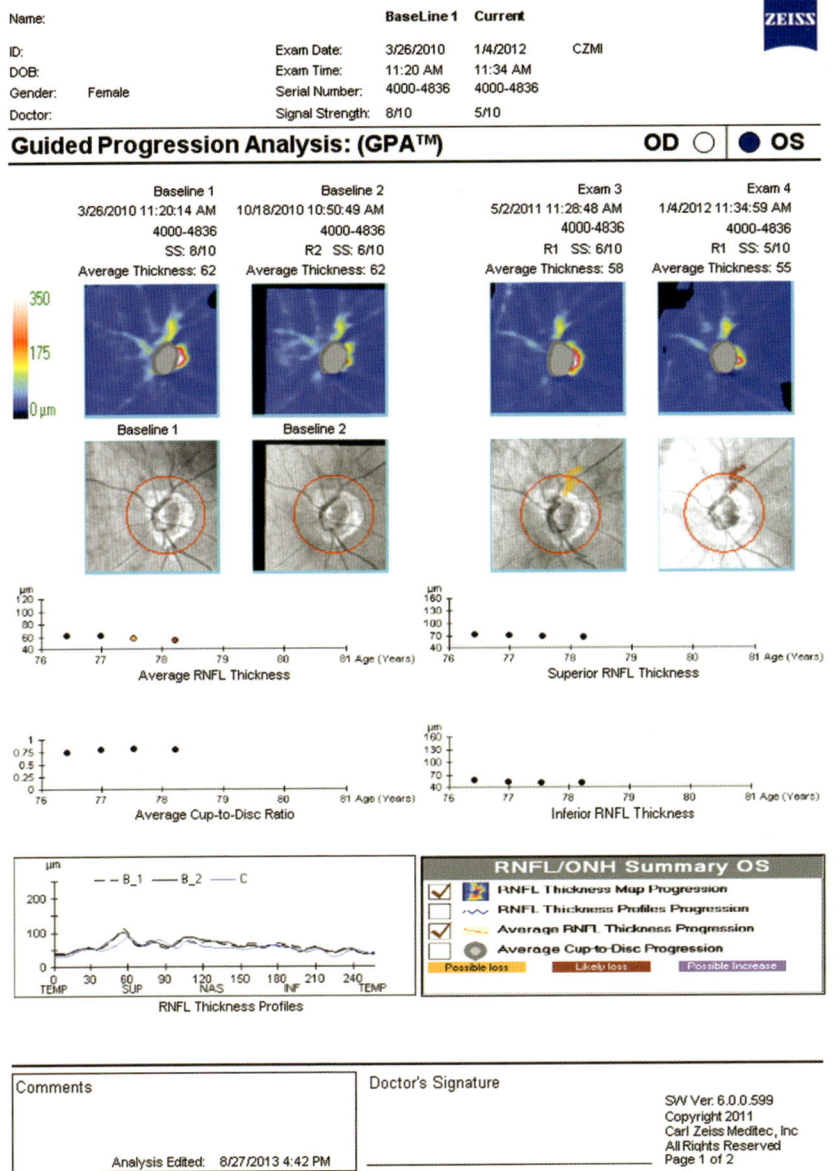

REFERENCES

1. Kotowski J, Wollstein G, Ishikawa H, Schuman JS. Imaging of the optic nerve and retinal nerve fiber layer: An essential part of glaucoma diagnosis and monitoring. Surv Ophthalmol. 2013 Oct 16.
2. Leung CK. Diagnosing progression with optical coherence tomography. Curr Opin Ophthalmol. 2013 Dec 23.
3. Quigley HA, Katz J, Derick RJ, Gilbert D, Sommer A. An evaluation of optic disc and nerve fiber layer examinations in monitoring progression of early glaucoma damage. Ophthalmology. 1992;99(1):19-28.
4. Na JH, Sung KR, Lee JR, Lee KS, Baek S, Kim HK, Sohn YH. Detection of glaucomatous progression by spectral-domain optical coherence tomography. Ophthalmology. 2013;120(7):1388-95.
5. Grewal DS, Tanna AP. Diagnosis of glaucoma and detection of glaucoma progression using spectral domain optical coherence tomography. Curr Opin Ophthalmol. 2013;24(2):150-61.
6. Anraku A, Enomoto N, Takeyama A, Ito H, Tomita G. Baseline thickness of macular ganglion cell complex predicts progression of visual field loss. Graefes Arch Clin Exp Ophthalmol. 2014;252(1):109-15.
7. Meira-Freitas D, Lisboa R, Tatham A, Zangwill LM, Weinreb RN, Girkin CA, Liebmann JM, Medeiros FA. Predicting progression in glaucoma suspects with longitudinal estimates of retinal ganglion cell counts. Invest Ophthalmol Vis Sci. 2013;54(6):4174-83.
8. Medeiros FA, Zangwill LM, Anderson DR, Liebmann JM, Girkin CA, Harwerth RS, Fredette MJ, Weinreb RN. Estimating the rate of retinal ganglion cell loss in glaucoma. Am J Ophthalmol. 2012;154(5):814-24.e1.
9. Lee JR, Sung KR, Na JH, Shon K, Lee KS. Discrepancy between optic disc and nerve fiber layer assessment and optical coherence tomography in detecting glaucomatous progression. Jpn J Ophthalmol. 2013;57(6):546-52.
10. Tenkumo K, Hirooka K, Baba T, Nitta E, Sato S, Shiraga F. Evaluation of relationship between retinal nerve fiber layer thickness progression and visual field progression in patients with glaucoma. Jpn J Ophthalmol. 2013 Jun 25.
11. Sehi M, Bhardwaj N, Chung YS, Greenfield DS. Advanced Imaging for Glaucoma Study Group. Evaluation of baseline structural factors for predicting glaucomatous visual-field progression using optical coherence tomography, scanning laser polarimetry and confocal scanning laser ophthalmoscopy. Eye (Lond). 2012;26(12):1527-35.

CHAPTER

7

Ganglion Cell Analysis

■ INTRODUCTION

Glaucoma is characterized by the loss of axons in the optic nerve head which results in the neuroretinal loss and increased cupping. The axon loss is due to ganglion cell dropout which is measurable by the optical coherence tomography (OCT). It has been shown to be a valid tool for glaucoma diagnostics.[1,2] Essentially, the width of the ganglion cell layer (ganglion cell-inner plexifom layer) is the prime parameter for diagnosis though novel ones such as the minimal single spoke thickness is also useful.[3,4] The ganglion cell thickness is shown to correlate with the retinal nerve fiber layer (RNFL) loss, visual field loss, contrast sensitivity function, cup:disc ratio and degree of relative afferent pupillary defect.[5-10] Ganglion cell layer analysis on the OCT has been shown to have good reproducibility and may be used for long-term follow-up.[11] In cases of parafoveal visual field loss, ganglion cell thickness measurement has shown better ability to diagnose glaucoma than RNFL changes.[11] However, other situations such as cataract tend to have a worse impact on the use of ganglion cell thickness over RNFL as a measure to diagnose glaucoma.[12]

The baseline ganglion cell layer thickness has been shown to predict visual field progression and literature also demonstrates that ganglion cells could be used to diagnose and determine progression.[13-15]

Imaging of the ganglion cell has thus far been indirect and the newer research in enhancing technology to enable direct visualization of ganglion cells is bearing fruit. In the future, direct visualization of these cells, their derivatives as well as apoptosis visualization may be commercially possible.[16]

The evaluation of ganglion cell loss is gaining increasing importance with newer insights into its role in possibly predicting long-term blindness in both glaucomatous and non-glaucomatous optic neuropathies as well as assessing the progression of various systemic neurologic diseases.

■ TECHNOLOGY

The ganglion cell analysis measures the thicknesses for the sum of the ganglion cell layer and inner plexiform layer (GCL + IPL layers) using data from the macular 200 × 200 or 512 × 128 cube scan patterns (Figure 7.1). Comparisons are made to normative data in a manner similar to that for the RNFL. This implies that the actual measurement is not the ganglion cell layer alone, though any changes in this layer would reflect on the GCL + IPL combination.

■ METHOD

The ganglion cell analysis is derived from the macular cube examination. The option to analyze the ganglion cell layer is available in OCT machines which have this particular module installed. The macular cube option is selected from the OCT scan options and the patient is asked to fixate at the central fixation target displayed on the patient's viewing screen. The operator will see an acquisition square on his screen which is then centred on the fovea using fine adjustments. The image is optimized and captured. This is followed by selecting the ganglion cell analysis option from the analysis screen. The output appears as is shown in Figure 7.2.

■ ANALYSIS

The ganglion cell analysis screen contains the following elements:
- *Thickness acquisition map:* It shows thickness measurements of the GCL + IPL in the 6 mm by 6 mm cube. It is displayed as an elliptical annulus centred about the fovea (Figure 7.3).

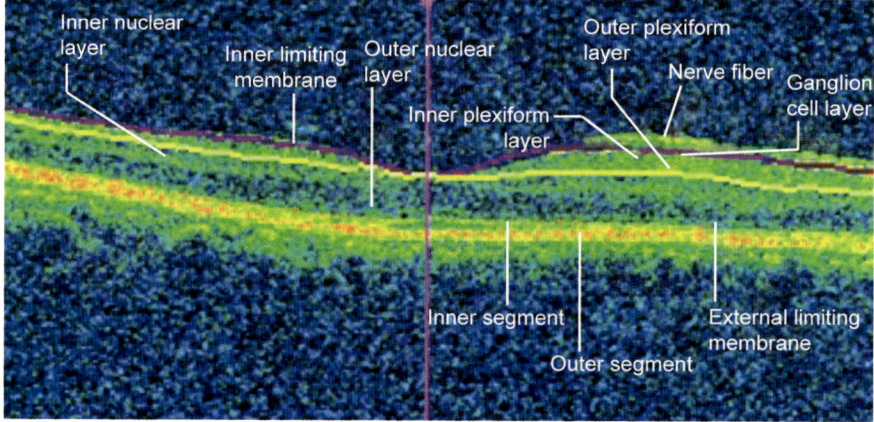

Fig. 7.1 Layers of the retina as seen on an OCT scan. Note the ganglion cell layer and the inner plexiform layer

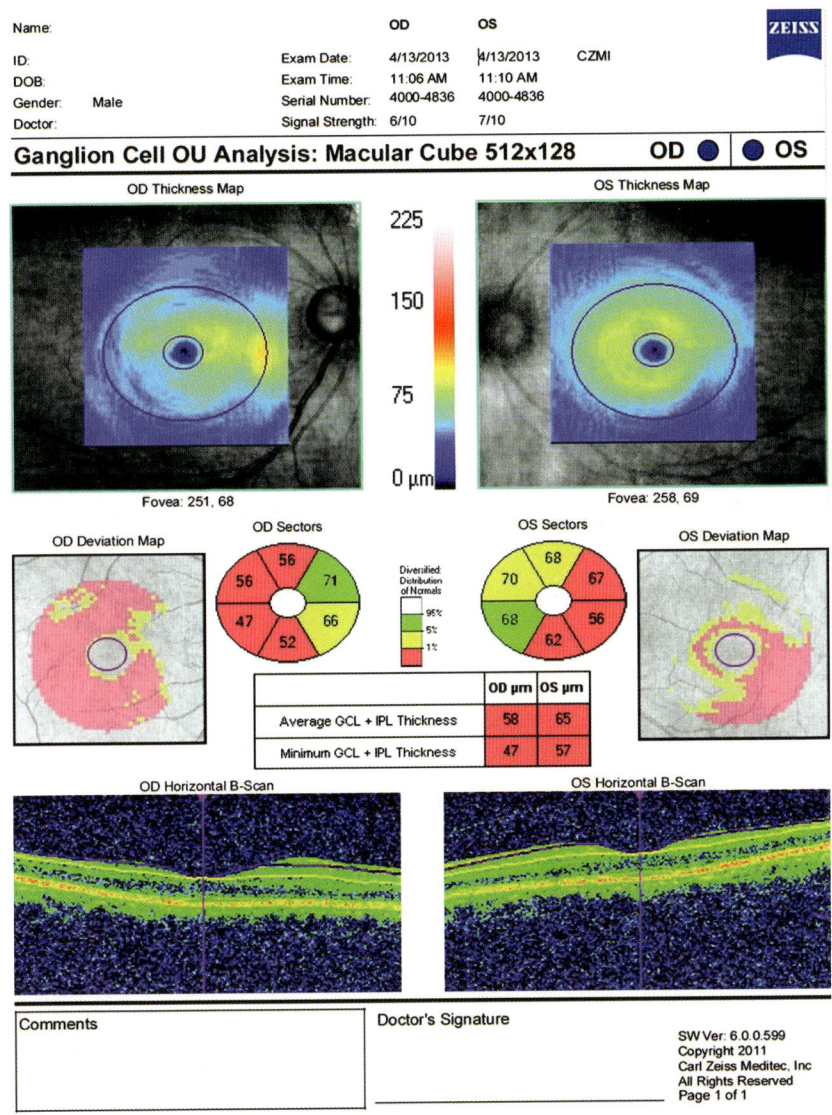

Fig. 7.2 The output of a ganglion cell analysis

- *Deviation map:* It shows a comparison of GCL + IPL thickness to normative data with red and yellow indicating that the thickness is expected to be normal in less than 1% and 5% of the population respectively. This is displayed in the form of red or yellow superpixels on the gray scale photograph (Figure 7.4).
- *Thickness table:* Showing average and minimum thicknesses within the elliptical annulus (Figure 7.5).

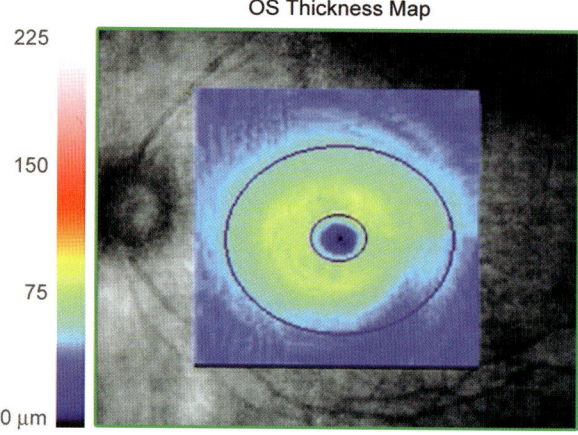

Fig. 7.3 Thickness acquisition map

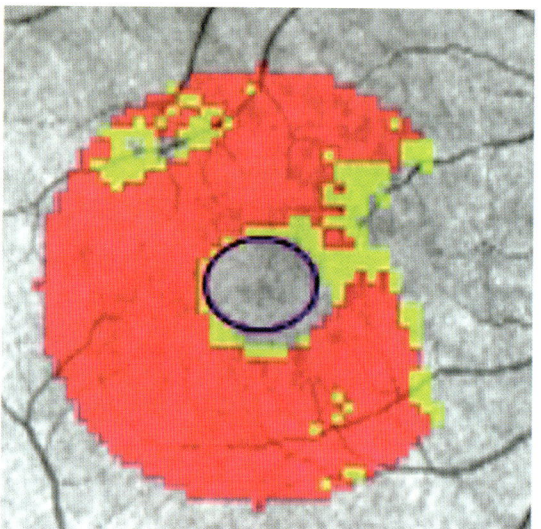

Fig. 7.4 Deviation map

	OD μm	OS μm
Average CGL + IPL thickness	58	65
Minimum CGL + IPL thickness	47	57

Fig. 7.5 Thickness table during a ganglion cell analysis

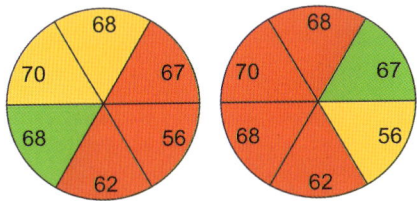

Fig. 7.6 Figure displaying sectoral thickness map of both eyes

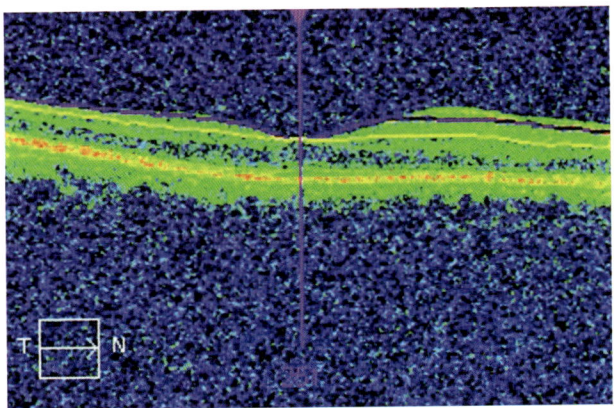

Fig. 7.7 Extracted B Scan

- *Sectoral thickness map:* Displayed in an elliptical manner divided into six sectors—three superior and three inferior, and are color coded in a manner similar to the RNFL thickness maps (Figure 7.6).
- *Horizontal and vertical B-scans:* These are extracted from the macular cube and the locales are marked on the macular map (Figure 7.7).

The analyzer allows editing the fovea location and navigating through the B-scans. If the fovea is not located, the foveal location and measurement circles are centered in the 6 mm square.

■ CONCLUSION

The ganglion cell analysis is a relatively new form of objective assessment of glaucomatous damage and holds good promise for predicting and assessing extent of axonal loss and possible blindness. However, our understanding of its interpretation and their implications is still in a moderately nascent stage and thus its use in routine clinical practice is still limited.

Key Points
- Ganglion cell layer thickness is sensitive to glaucoma related changes.
- Ganglion cell analysis is an indirect measure of ganglion cell dropout.
- Ganglion cell analysis is more sensitive for paracentral scotomas.
- Retinal disorders affecting the architecture of retinal layers should be ruled out before ganglion cell analysis.

CASES

CASE 1: UNDERSTANDING THE GANGLION CELL ANALYSIS. ALWAYS COMPARE WITH RNFL ANALYSIS

This is a case of a patient having advance unilateral glaucoma in the right eye and a normal optic nerve in the left eye. Notice the advance cupping in the right eye in the gray-scale fundus photograph in the acquisition thickness map of the ganglion cell analysis. Also note that the ganglion cell thickness map is showing a blue coloration suggestive of extreme thinning which is evident in the large areas of red superpixels on the deviation map. The RNFL map of the same patient is showing severe thinning in the superior, temporal and inferior quadrants while the nasal quadrant appears normal. This apparent mismatch between the two scans is due to the fact that the ganglion cell analysis is probably more sensitive than the RNFL change and has captured axonal loss earlier than the RNFL map. The left eye is normal.

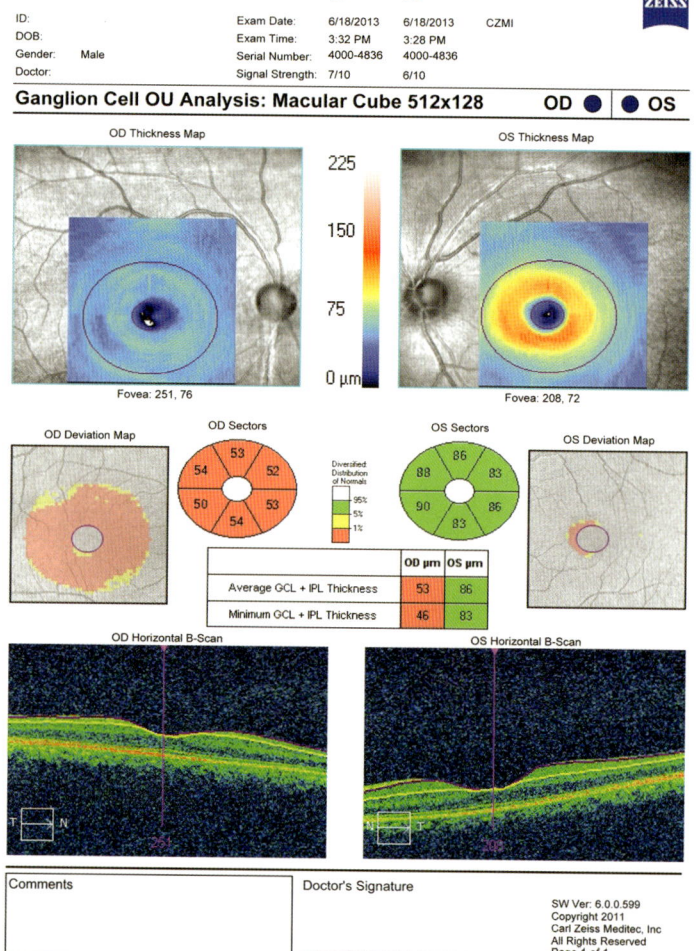

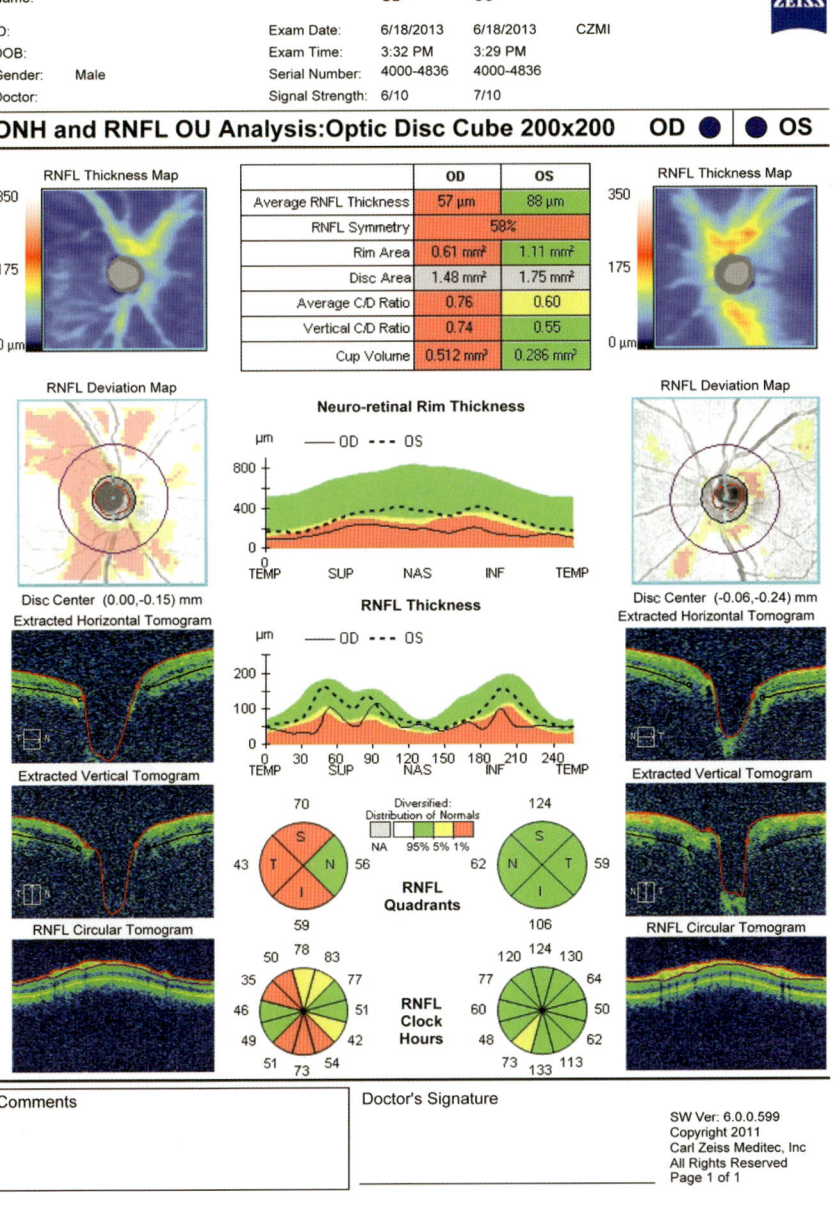

CASE 2: ADVANCED GLAUCOMA

This is a patient with advanced glaucomatous neuropathy in both eyes showing severe thinning on the ganglion cell analysis along with thinning on the RNFL analysis. Note the bunting of the RNFL peaks on the Temporal-superior-nasal-inferior-temporal (TSNIT)-RNFL profile. This case exemplifies the strong correlation between ganglion cell loss and RNFL loss.

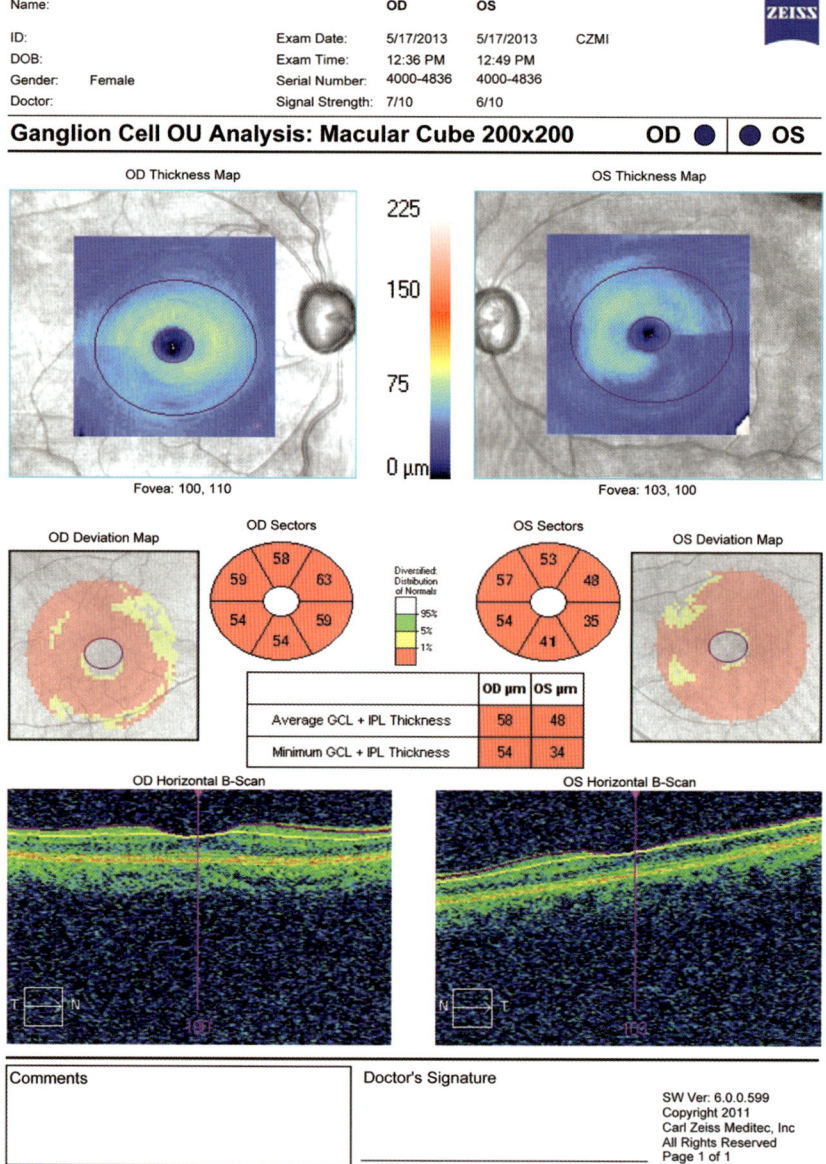

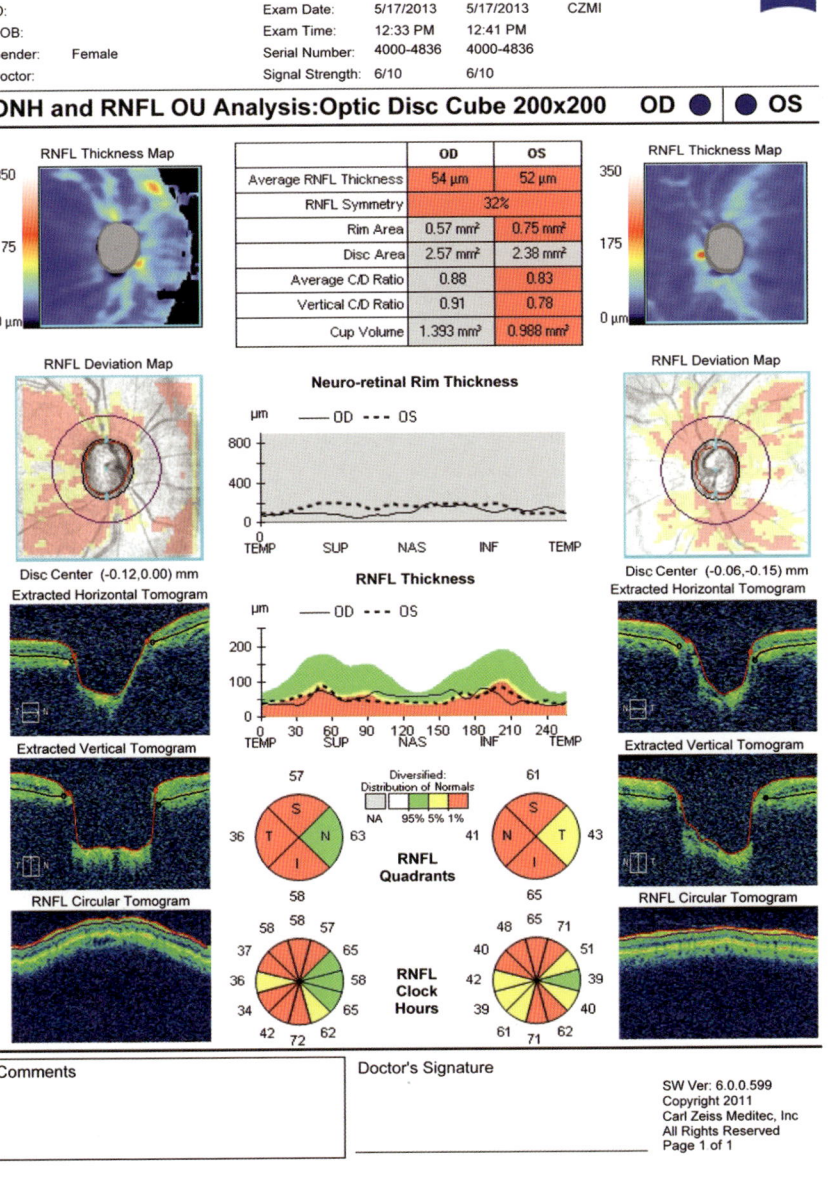

CASE 3: PREPERIMETRIC AND PERIMETRIC GLAUCOMA

This is a middle-aged patient with a typical glaucomatous visual field defect in the right eye (perimetric glaucoma) and a normal field in the left eye (preperimetric glaucoma). There is moderate cupping in both eyes with RNFL thinning of the superotemporal RNFL in the right eye and inferotemporal quadrant of the left eye. Note that there is corresponding thinning on the ganglion cell layer analysis indicating that it also can help to detect preperimetric glaucoma.

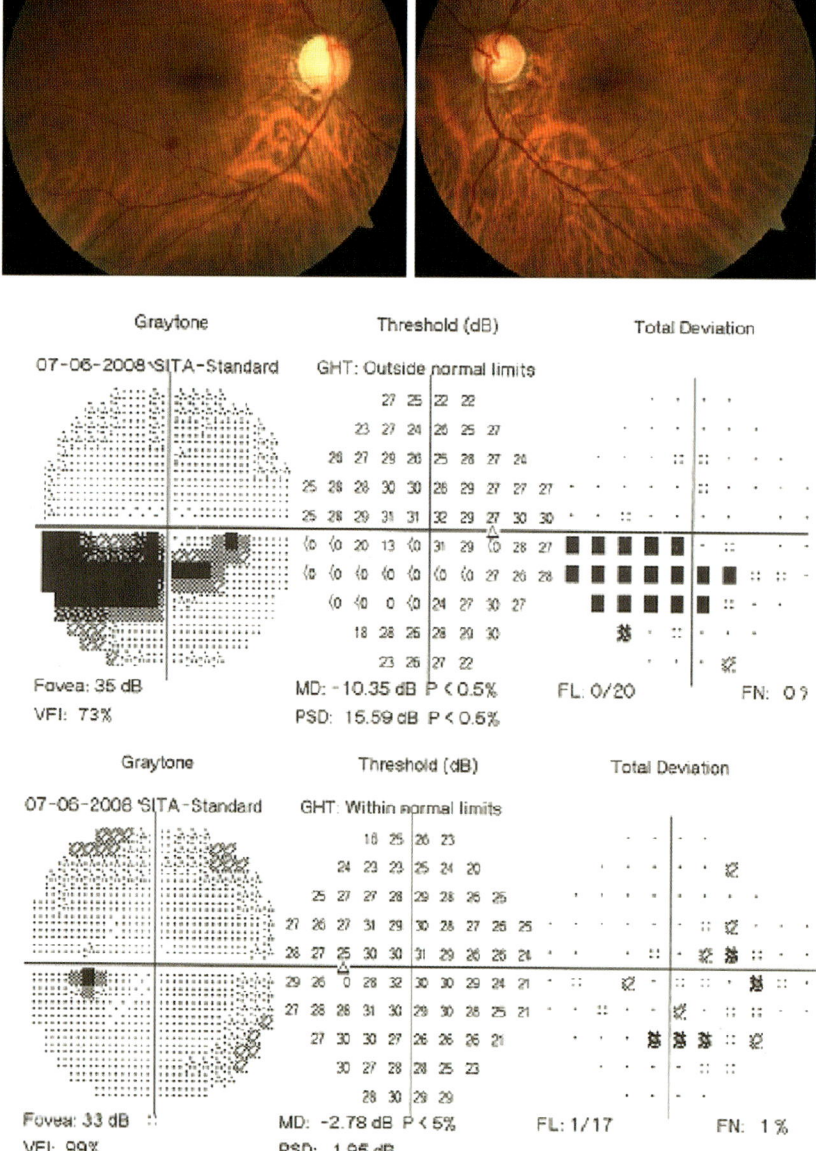

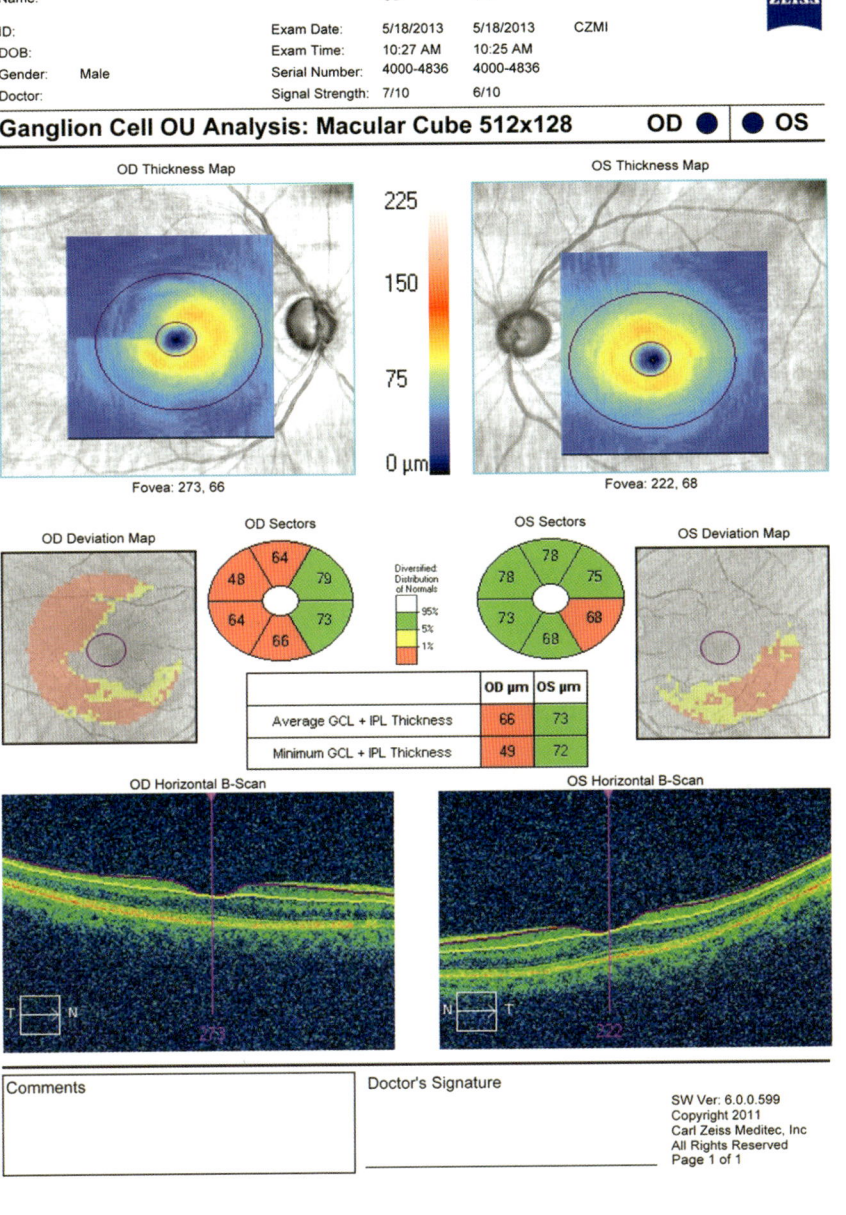

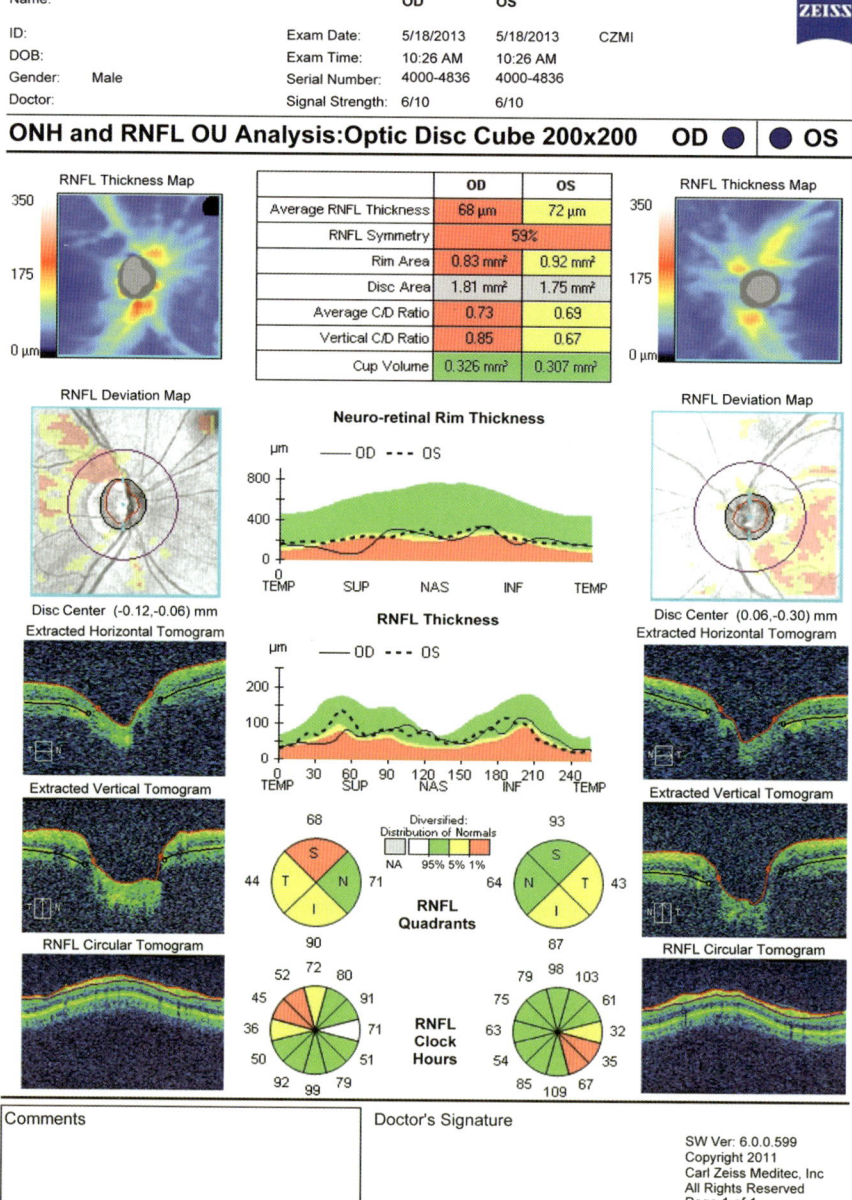

REFERENCES

1. Sung MS, Yoon JH, Park SW. Diagnostic validity of macular ganglion cell: inner Plexiform Layer Thickness Deviation Map Algorithm Using Cirrus HD-OCT in Preperimetric and Early Glaucoma. J Glaucoma. 2013.
2. Nouri-Mahdavi K, Nowroozizadeh S, Nassiri N, Cirineo N, Knipping S, Giaconi J, Caprioli J. Macular ganglion cell/inner plexiform layer measurements by spectral domain optical coherence tomography for detection of early glaucoma and comparison to retinal nerve fiber layer measurements. Am J Ophthalmol. 2013;156(6):1297-307.e2.
3. Rimayanti U, Latief MA, Arintawati P, Akita T, Tanaka J, Kiuchi Y. Width of abnormal ganglion cell complex area determined using optical coherence tomography to predict glaucoma. Jpn J Ophthalmol. 2014;58(1):47-55.
4. Takayama K, Hangai M, Durbin M, Nakano N, Morooka S, Akagi T, Ikeda HO, Yoshimura N. A novel method to detect local ganglion cell loss in early glaucoma using spectral-domain optical coherence tomography. Invest Ophthalmol Vis Sci. 2012;53(11):6904-13.
5. Tatham AJ, Weinreb RN, Zangwill LM, Liebmann JM, Girkin CA, Medeiros FA. Estimated retinal ganglion cell counts in glaucomatous eyes with localized retinal nerve fiber layer defects. Am J Ophthalmol. 2013;156(3):578-87.e1.
6. Mwanza JC, Budenz DL, Godfrey DG, Neelakantan A, Sayyad FE, Chang RT, Lee RK. Diagnostic performance of optical coherence tomography ganglion cell-inner plexiform layer thickness measurements in early glaucoma. Ophthalmology. 2014.
7. Le PV, Tan O, Chopra V, Francis BA, Ragab O, Varma R, Huang D. Regional correlation among ganglion cell complex, nerve fiber layer, and visual field loss in glaucoma. Invest Ophthalmol Vis Sci. 2013;54(6):4287-95.
8. Tatham A, Meira-Freitas D, Weinreb RN, Marvasti AH, Zangwill LM, Medeiros FA. Estimation of retinal ganglion cell loss in glaucomatous eyes with a relative afferent pupillary defect. Invest Ophthalmol Vis Sci. 2013 Nov 26. pii: iovs.13-12921v1.
9. Tatham AJ, Weinreb RN, Zangwill LM, Liebmann JM, Girkin CA, Medeiros FA. The relationship between cup-to-disc ratio and estimated number of retinal ganglion cells. Invest Ophthalmol Vis Sci. 2013;54(5):3205-14.
10. Francoz M, Fenolland JR, Giraud JM, El Chehab H, Sendon D, May F, Renard JP. Reproducibility of macular ganglion cell-inner plexiform layer thickness measurement with cirrus HD-OCT in normal, hypertensive and glaucomatous eyes. Br J Ophthalmol. 2013.
11. Shin HY, Park HY, Jung KI, Choi JA, Park CK. Glaucoma diagnostic ability of ganglion cell-inner plexiform layer thickness differs according to the location of visual field loss. Ophthalmology. 2014;121(1):93-9.
12. Nakatani Y, Higashide T, Ohkubo S, Takeda H, Sugiyama K. Effect of cataract and its removal on ganglion cell complex thickness and peripapillary retinal nerve fiber layer thickness measurements by fourier-domain optical coherence tomography. J Glaucoma. 2013;22(6):447-55.
13. Medeiros FA, Zangwill LM, Anderson DR, Liebmann JM, Girkin CA, Harwerth RS, Fredette MJ, Weinreb RN. Estimating the rate of retinal ganglion cell loss in glaucoma. Am J Ophthalmol. 2012;154(5):814-24.e1.
14. Anraku A, Enomoto N, Takeyama A, Ito H, Tomita G. Baseline thickness of macular ganglion cell complex predicts progression of visual field loss. Graefes Arch Clin Exp Ophthalmol. 2014;252(1):109-15.
15. Meira-Freitas D, Lisboa R, Tatham A, Zangwill LM, Weinreb RN, Girkin CA, Liebmann JM, Medeiros FA. Predicting progression in glaucoma suspects with longitudinal estimates of retinal ganglion cell counts. Invest Ophthalmol Vis Sci. 2013;54(6):4174-83.
16. Werkmeister RM, Cherecheanu AP, Garhofer G, Schmidl D, Schmetterer L. Imaging of retinal ganglion cells in glaucoma: pitfalls and challenges. Cell Tissue Res. 2013;353(2):261-8.

CHAPTER

8

Macular OCT

■ INTRODUCTION

The macula has always been an area of attention during the development of the optical coherence tomography (OCT) and imaging the macula was among the earliest uses of this machine in clinical practice. While the macular OCT is of utmost importance for retinal disorders, it is also of interest to glaucoma specialists. There are specific indications for examining the macula in a case of glaucoma. These include:
- Evaluating the retinal nerve fiber layer (RNFL) thickness at the macula and its progression (Figures 8.1 and 8.2)
- Evaluating for cystoid macular edema after a laser or surgical procedure or in patients who have been receiving prostaglandins (Figure 8.3)
- Evaluating hypotonic maculopathy after trabeculectomy and following up the patient for improvement (Figure 8.4).
- Ganglion cell analysis
- Useful for documenting concomitant retinal pathologies in glaucoma patients

Macular thinning has been shown to predict glaucoma and aids in its diagnosis, especially if it is asymmetrical macular thinning.[1] The central macular thickness has shown better predictability for central or paracentral vision loss than central RNFL in glaucoma patients.[2] In fact, macular thinning is predictive of visual function loss in glaucoma.[3] In contrast to macular thinning, an increase in thickness has been noted in the post-trabeculectomy period reflective of a hypotony and possible subclinical inflammation.[4]

■ TECHNOLOGY

The macular scan can be captured either in a macular 200 × 200 or 512 × 128 cube pattern or in a 5 line raster pattern. The macular cube is required for measurement of macular thickness scans and ganglion cell analysis while the raster helps define pathology such as cystoid macular edema or hypotonic maculopathy. The raster scans can be used to measure macular thickness at any point along the scan but

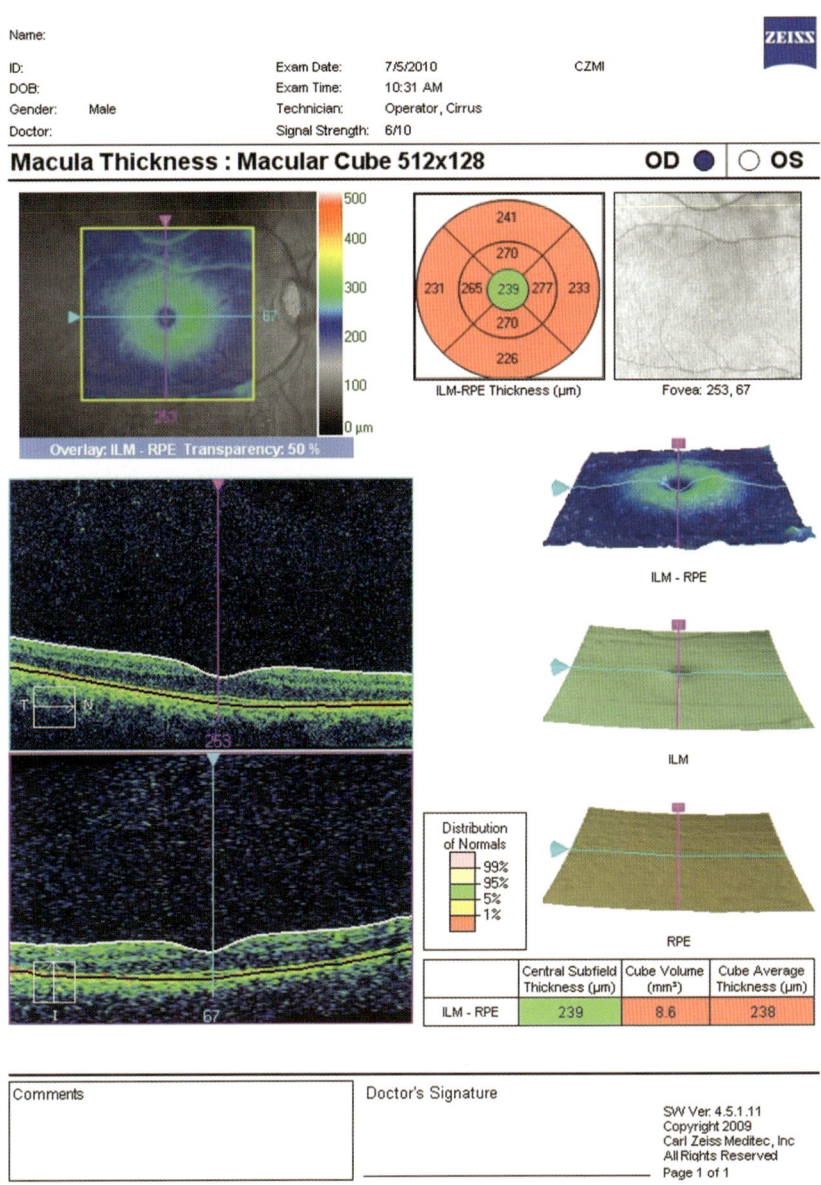

Fig. 8.1 Printout of a macular cube analysis showing thinning of the macula in a case of advanced glaucoma. The macular scan is divided into three zones, the inner circle represents an area of the fovea (central 1 mm), the middle circle represents the inner macula (1-3 mm from fovea) and the outer circle represents the outer macula (3-5 mm diameter). The normal macular thickness as measured by the Stratus OCT is around 200 μm while by the Cirrus OCT, it is around 260 μm

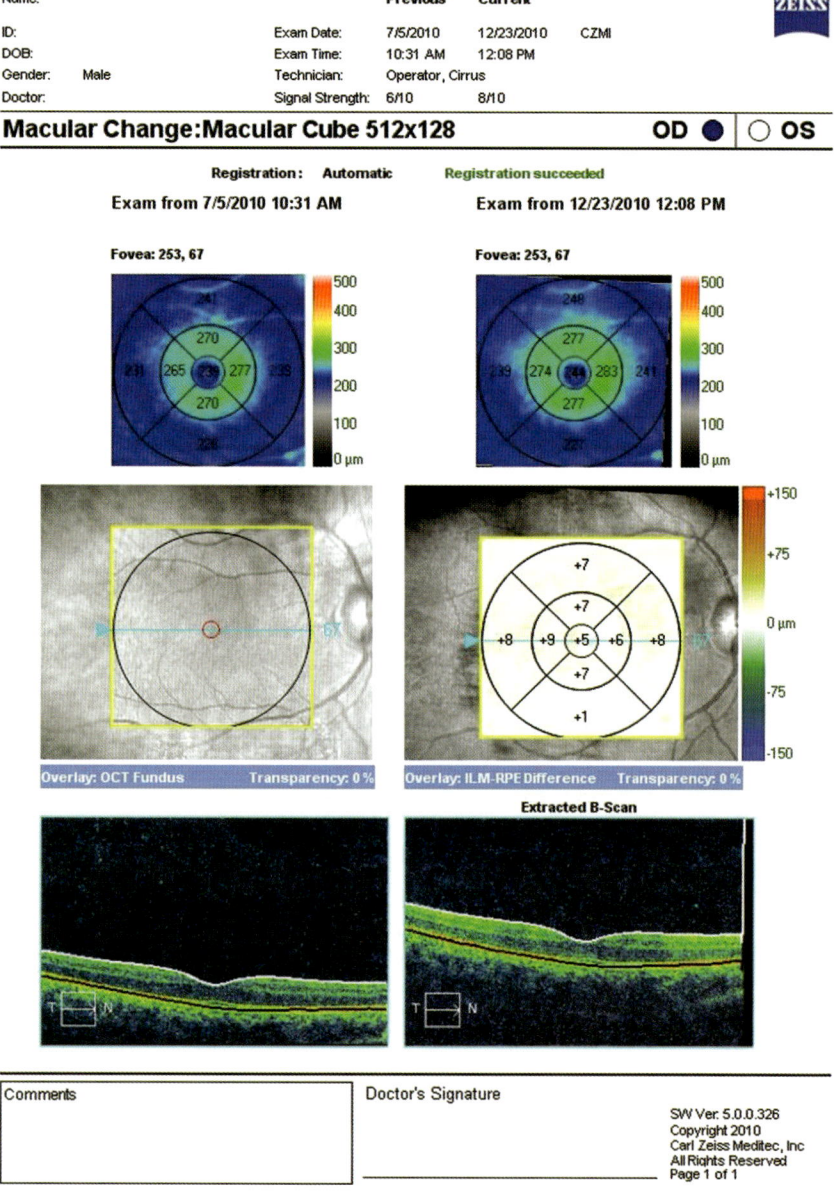

Fig. 8.2 A macular change analysis done to evaluate progression. Note the minimally increased macular thickness shown in the form of + signs on the progression analysis

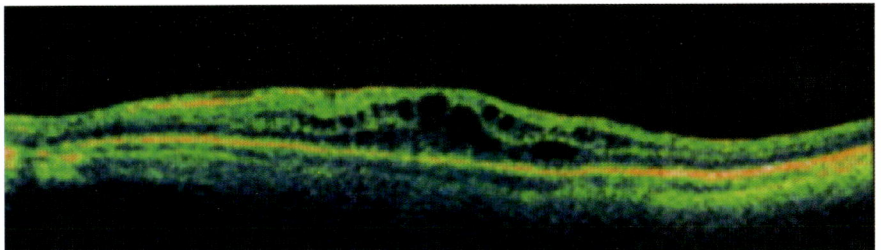

Fig. 8.3 A raster scan demonstrating cystoid macular edema

do not give automated values unlike the macular cube. Just as in any OCT scan, the 5-line raster displays the retina in hyper-reflective and hyporeflective layers. High reflectivity structures in the retina include the retinal nerve fiber layer, the plexiform layers and the retinal pigment epithelium which are displayed in warm colors such as red while low reflectivity structures which include the ganglion cell layer, the nuclear layers and the major retinal vessels and are displayed in cool colors such as blue and black (Figure 8.4).

■ METHOD

The macular scans are best done with a dilated pupil though a 3 cmm pupil can permit an accurate scan in the newer generation OCT. The operator should select a macular cube option where by a square shaped acquisition area is focused on the macula and the scan captured after optimizing. This scan provides limited anatomical details but can be renedered into a 3D image to give a birds-eye view of the macula. Alternatively an operator may choose the five line raster scan which consists of five parallel B-scans the orientation of which can be changed between vertical, horizontal or at any desired angle. The spacing between the lines can be increased or decreased. These lines can be dragged and placed at any area on the displayed fundus image on the OCT console to give a localized anatomy. The scans help define the anatomy very well and manual calculations of thickness can be done using a calliper.

■ ANALYSIS

The basic derivation from the macular scan which is of use to glaucoma specialists is the macular thickness attained from a macular cube scan. The other important parameter, namely the ganglion cell thickness, has been previously detailed.

In the analysis of the macular cube scan, the important parameter to look for is the macular thickness at the parafoveal area as it is this thickness and its asymmetry in the superior versus inferior manner that has been shown to be predictive of glaucoma. The thickness is compared to the normative database in a manner similar to the RNFL and the significantly abnormal areas at 5% and 1% levels are marked yellow and red respectively.

The interpretation of the 5-line raster is qualitative rather than quantitative and is useful while diagnosing cystoid macular edema and hypotonic maculopathy.

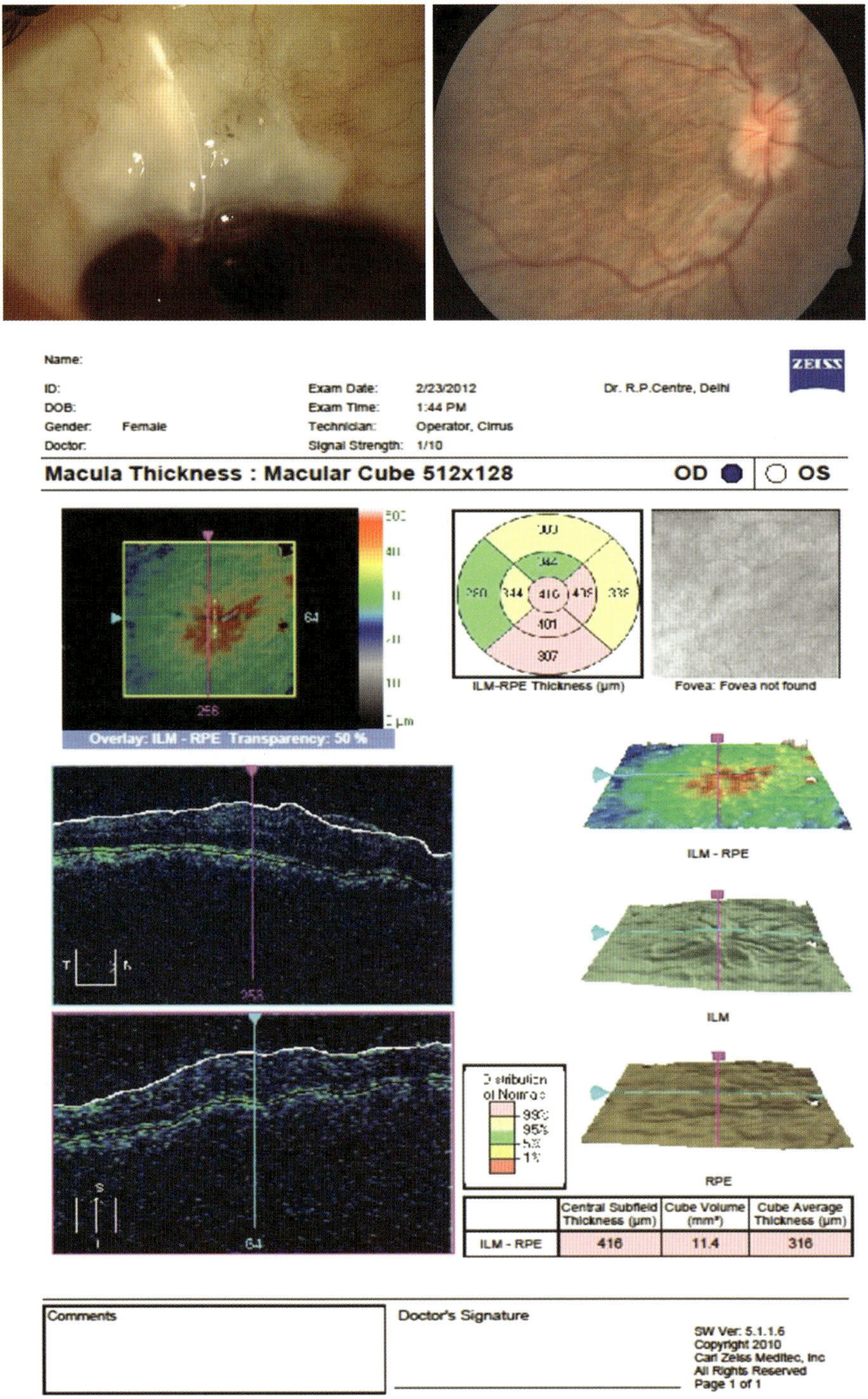

Fig. 8.4 An eye after trabeculectomy presenting with hypotonic maculopathy. Note the loss of foveal contour on OCT and the thickening of macula (edema)

In the former, cystoid spaces would be seen in the foveal and perifoveal region with increased thickness while in the latter, a spongy thickness of the macula is noted. A more detailed interpretation of the macular OCT is for retinal conditions and beyond the scope of this book.

> **Key Points**
> - Macular OCT is helpful in glaucoma for enabling ganglion cell analysis, diagnosing cystoid macular edema and hypotonic maculopathy.
> - It is possible to suspect glaucoma on the basis of an abnormal macular cube scan, however it does not form the mainstay of diagnosis.

REFERENCES

1. Lee KS, Lee JR, Na JH, Kook MS. Usefulness of macular thickness derived from spectral-domain optical coherence tomography in the detection of glaucoma progression. Invest Ophthalmol Vis Sci. 2013;54(3):1941-9.
2. Sullivan-Mee M, Ruegg CC, Pensyl D, Halverson K, Qualls C. Diagnostic precision of retinal nerve fiber layer and macular thickness asymmetry parameters for identifying early primary open-angle glaucoma. Am J Ophthalmol. 2013;156(3):567-77.e1
3. Agrawal S, Singh V, Bhasker SK, Sharma B. Correlation of visual functions with macular thickness in primary open angle glaucoma. Oman J Ophthalmol. 2013;6(2):96-8.
4. Sesar A, Cavar I, Sesar AP, Geber MZ, Sesar I, Laus KN, Vatavuk Z, Mandić Z. Macular thickness after glaucoma filtration surgery. Coll Antropol. 2013;37(3):841-5.

CHAPTER

9

Anterior Segment Analysis

■ INTRODUCTION

Anterior segment imaging has been another important aspect of imaging a case of glaucoma. In the past this has been done in the simplest manner using a gonioscope and a slit lamp mounted camera to view the angle. The ultrasound biomicroscopy helped us gain access to the hitherto unimaged areas of the eye though the resolution and the quantitative aspect of imaging were not perfect. The posterior segment OCT gave great images of the retina with good quantification and a high resolution and was adapted to image and analyse the angle.[1] Currently this task is done with the anterior segment OCT which is discussed in chapter 3. The Cirrus OCT anterior segment module has been used in studies to successfully locate the Schwalbe's line in eyes for enabling placement of glaucoma drainage implants in the future.[2]

■ TECHNOLOGY AND USE

The Cirrus OCT contains a module for acquiring anterior segment scans. These are in the form of either an Anterior Segment Cube 512 × 128 or Anterior Segment 5 Line Raster. Anterior Segment Cube 512 × 128 (Figure 9.1).

This scan uses a 4 millimeter square grid and can be used to evaluate the central corneal thickness and create a 3D image of the iris, angle and cornea (Figures 9.1 and 9.3). For the anterior segment scan, the OCT beam scans in an arc to allow the curved cornea to better fit into the 2 mm scan depth. Therefore, the cornea appears flat in the display during alignment and acquisition. This effect is partially corrected after the acquisition, so the cornea will finally appear with the expected curvature during review and analysis. The technique provides an excellent anatomic correlation though quantification is not as defined as for the macular cube.

The anterior segment 5-line raster mode scans through 5 parallel lines of equal length and can generate high resolution images of the anterior chamber angle and cornea (Figure 9.2). Each line is 3 mm long and contains 4096 A-scans. The Anterior Segment 5 line raster scan is the preferred scan type for imaging the anterior chamber angle because it can be rotated to image a cross section perpendicular to the limbus at any location. The iris is not well focused when scanning the angle. This gives a good anatomic definition of the angle and can be done in a dark room to give the true angle (Figure 9.4). The anterior segment 5-line raster can be used to view the bulbar conjunctiva including the bleb post a trabeculectomy surgery[3] (Figure 9.5).

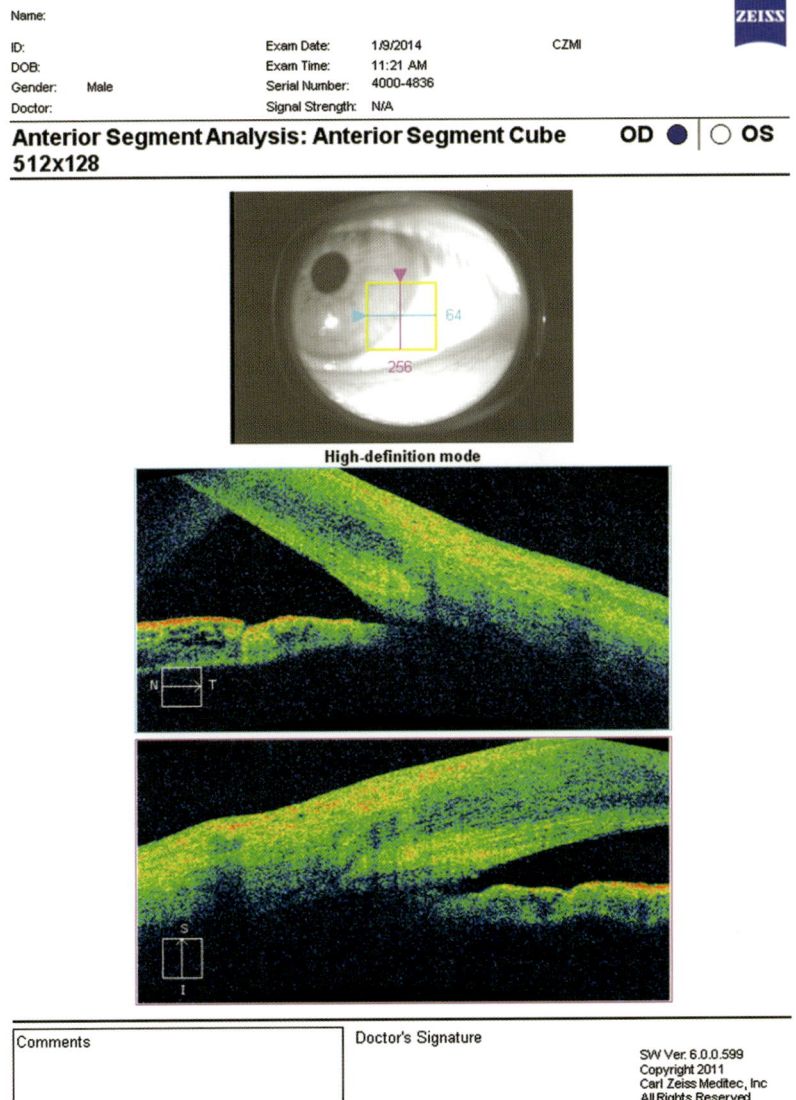

Fig. 9.1 The acquisition of the anterior segment images using the anterior segment cube program. The image is showing an open angle

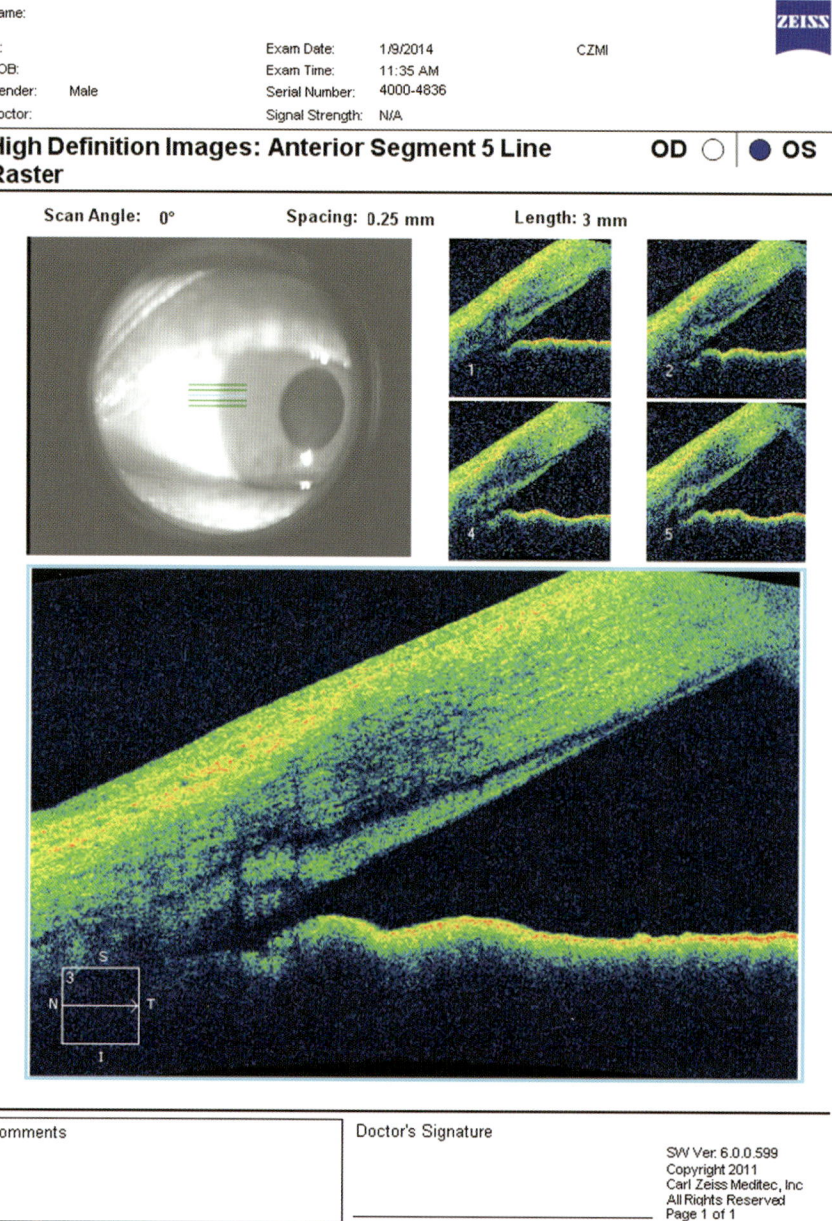

Fig. 9.2 The acquisition of the angle images using the anterior segment 5-line raster program. The image is showing a narrow angle

Anterior Segment Analysis 109

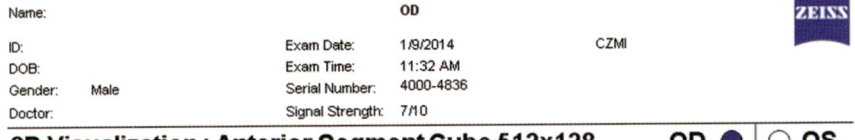

3D Visualization : Anterior Segment Cube 512x128   OD ● ○ OS

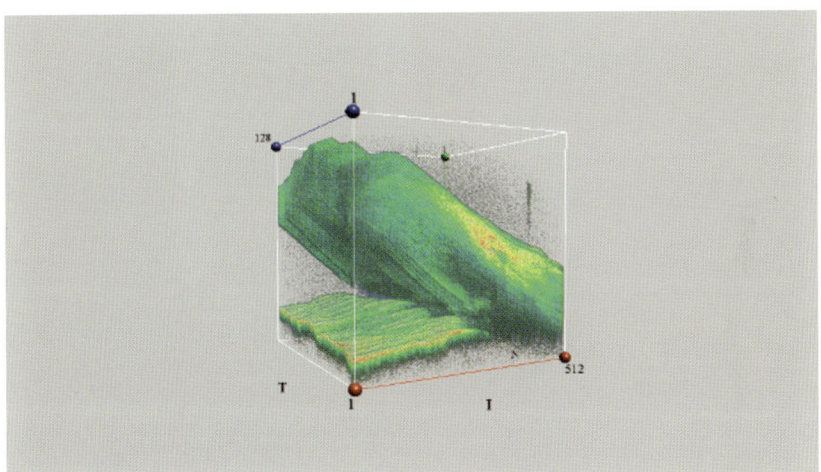

Brightness: 0
Contrast: 100
Threshold: 40
Transparency (%): 0
Use Same Transparency for all Pixels: False

Apply Intensity Filter: False
Intensity Value: 0
Intensity Range: 0
Surface Light Weight: 25
Gradient Step Size: 5
Lighting Enabled: True

Cube cut at Width: 1, 512 Height: 1, 128 A-Scan: 1, 1024

Comments

Doctor's Signature

SW Ver. 6.0.0.599
Copyright 2011
Carl Zeiss Meditec, Inc
All Rights Reserved
Page 1 of 1

Fig. 9.3 The 3-D reconstruction of the angle after imaging the anterior segment using the anterior segment cube program

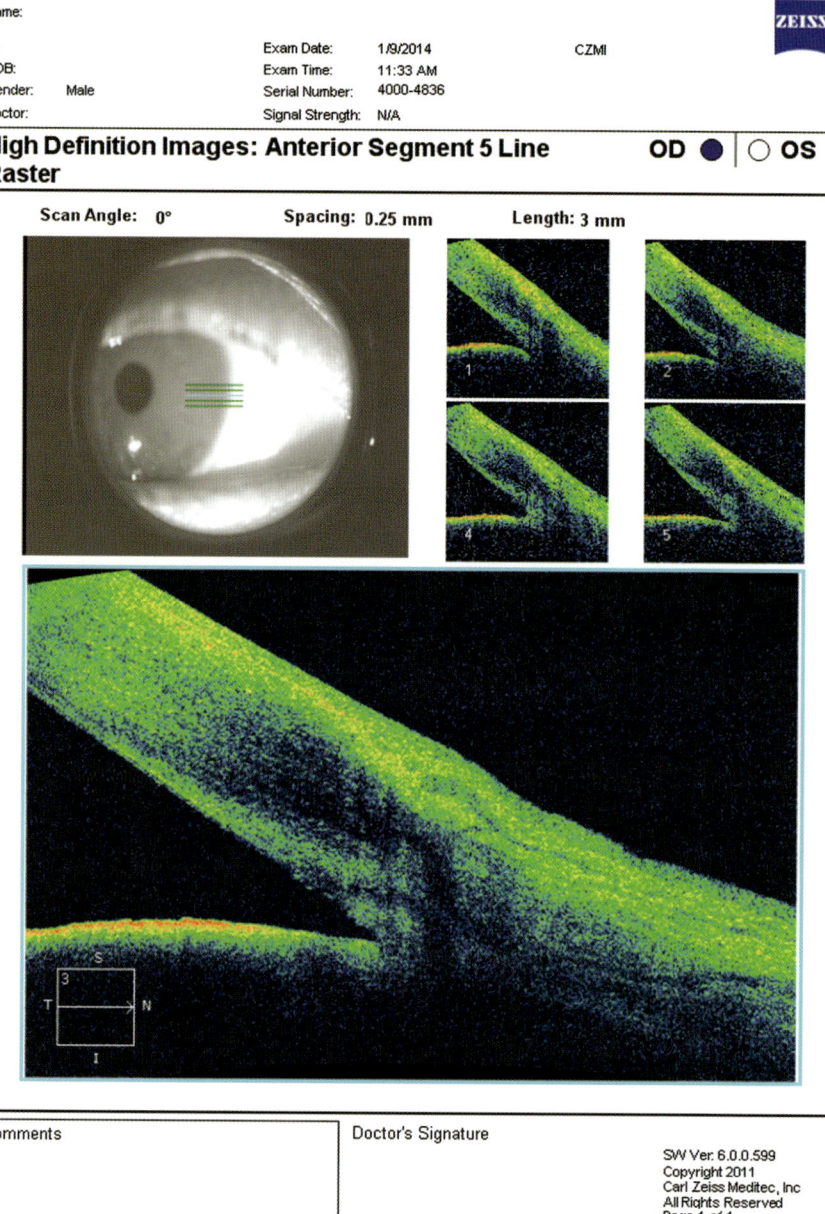

Fig. 9.4 The report of an anterior segment 5-line raster analysis as seen with a nasal to temporal scanning unlike Figure 9.2 which is acquired in a temporal to nasal manner

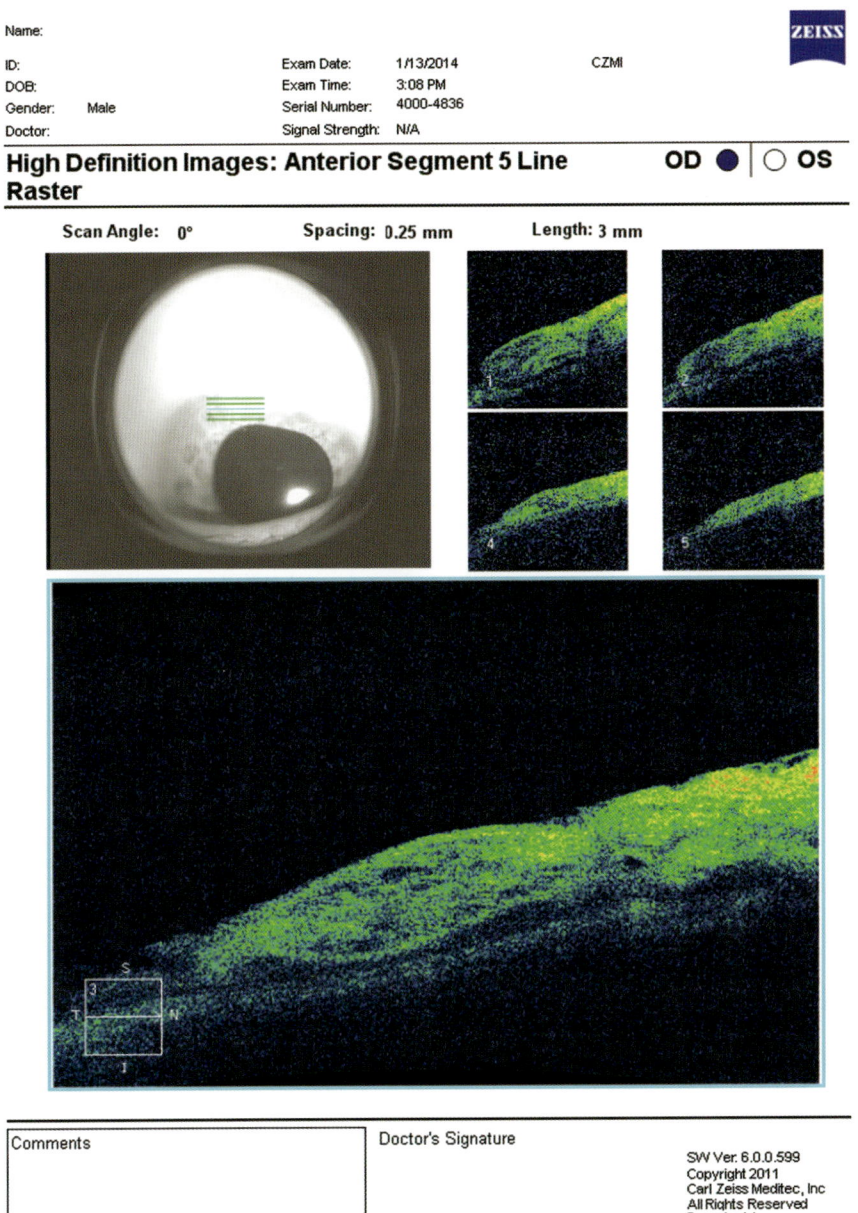

Fig. 9.5 Imaging of the bleb using a 5-line raster scan. Notice the fluid filled space below the bulbar conjunctiva of the bleb

CONCLUSION

The posterior segment OCT can be used to perform basic anterior segment imaging. It can evaluate whether the angle is open or closed after identification of the scleral spur. However it is limited in its scope as detailed angle measurements cannot be performed and simultaneous imaging of temporal and nasal angle is not possible. However this is limited in its scope and is of diminishing value with the advent of the anterior segment OCT.

Key Points

- Posterior segment OCT may be used to non-invasively image the cornea and the anterior chamber angle (digital gonioscopy).
- The anterior segment 5-line raster is the preferred module for anterior segment evaluation using the posterior segment OCT.

REFERENCES

1. Rodrigues EB, Johanson M, Penha FM. Anterior segment tomography with the cirrus optical coherence tomography. J Ophthalmol. 2012;2012:806-989.
2. Breazzano MP, Fikhman M, Abraham JL, Barker-Griffith AE. Analysis of Schwalbe's Line (Limbal Smooth Zone) by Scanning Electron Microscopy and Optical Coherence Tomography in Human Eye Bank Eyes. J Ophthalmic Vis Res. 2013;8(1):9-16.
3. Zhang X, Li Q, Liu B, Zhou H, Wang H, Zhang Z, Xiang M, Han Z, Zou H. In vivo cross-sectional observation and thickness measurement of bulbar conjunctiva using optical coherence tomography. Invest Ophthalmol Vis Sci. 2011;52(10):7787-91.

CHAPTER

10

Limitations of OCT and Imaging Artifacts

■ INTRODUCTION

Technology of any kind has got its strengths and weaknesses. The ability to gain maximum from a technology is limited by the understanding of the operator and the interpreter of using the technology. The OCT is no different and by understanding a few basic principles, one can acquire a good scan and avoid misinterpretation.

■ ACQUIRING A GOOD SCAN: PEARLS

The OCT is dependent on EM waves entering the eye and bouncing back from the retina or disc. This implies that any hurdle in the path of the wave would cause a reduction in proper signal and possibly fallacious scans. On this basis, it is easy to understand that ocular pathologies such as poor ocular surface, dry eye, corneal opacification, corneal edema, lenticular opacification, posterior capsular opacification and vitreous haze can cause a reduced signal and consequential poor scan quality. Poor quality scans are shown to adversely affect results.[1] Similarly an uncalibrated machine or one with dirty optics would lead to poor signal strength. Machines having a high noise in the acquired image also hamper the final interpretation.

In view of all this, the following steps would ensure a good quality scan and reduce errors:
- The patient should be seated in a comfortable stool/chair and the height of the OCT machine should be so adjusted such that the patient need not slouch or sit too upright. (This will prevent movement artifacts and make the imaging stress free).
- The Chin should be placed firmly on the chin rest and the forehead should touch the forehead band. This ensures there is minimal movement and the machine is able to focus well. Also it reduces chances of head tilt which is known to cause artifactual values.

114 Optical Coherence Tomography in Current Glaucoma Practice

- The ocular surface should not be dry (if so ask the patient to blink a few times or put a drop of lubricant) and the patient is asked to blink while the examiner is focussing the retina or disc.
- The patient is asked to focus on the fixation star displayed to enable centering of the acquisition area. If there is any pupillary axis opacity, the axis can be tilted or the image acquired eccentrically. For this, the patient may be asked to look slightly away from the fixation star or the external fixation light can be used.
- If the patient has poor fixation in the eye from which the image is to be acquired, the better eye can be used to fixate using an external fixation target.
- The refractive error correction of the patient should be incorporated into the focus to enable a sharp image.
- The image acquisition circle or grid should be centered on the area of interest and the viewing window on the screen should be seen to ensure the full image of the RNFL is visible without the top or bottom getting truncated (Figure 10.1).
- After proper placement, the image is optimized once.

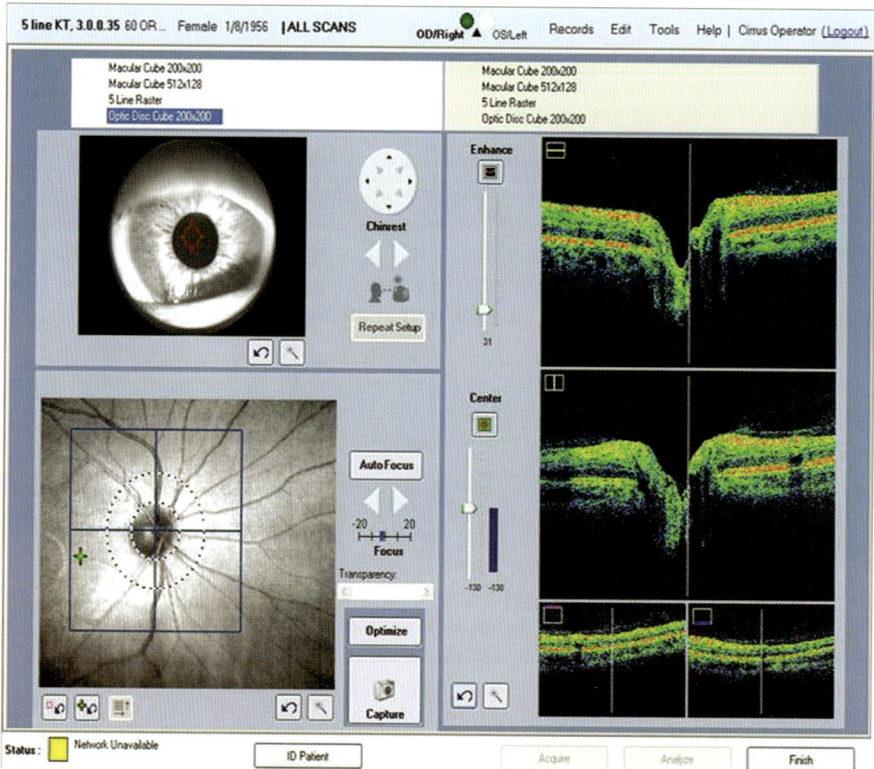

Fig. 10.1 Correct method of scan acquisition: A well-focused scan with the entire RNFL in the scan area

- Just prior to acquisition of the scan, the patient is asked to blink once and the image is then captured.
- After acquiring the image, the signal strength is seen prior to saving the image. The ideal signal strength should be 8 or more. However, a signal strength of 6 or more may be used for interpretation. It is important to remember that variations in signal strength may induce problems when interpreting progression of glaucomatous neuropathy on serial analysis.
- If a scan shows any abnormality, always reacquire the image and correlate with the fundus and visual fields before interpreting the scan.

PITFALLS IN OCT IMAGING

There may be various pitfalls in OCT image interpretation. This misinterpretation could depend on various factors.

Machine Factors

- An *improperly calibrated machine* leads to erroneous quantification of measurements
- Use of the machine in an *environment with excessive temperature fluctuations* or humidity could lead to imaging errors.
- *Deterioration of the optics* of the OCT may lead to a poor quality scan.

Scan Factors

- An improperly acquired scan with a *poor signal strength* or a low signal : noise ratio can provide wrong quantification measurements
- A scan which is decentered will not be useful to follow-up as the values obtained would be different with different amounts of *decentration*.
- A scan where the RNFL layer is truncated at the top or bottom due to *improper focussing* would lead to erroneously low values of RNFL (possibly even 0) (Figure 10.2).

Patient Factors

- *Poor fixation* of the patient could lead to segmentation of the scan as the machine is unable to track the saccades or microsaccades.
- *Head tilt* can lead to variable readings in the same area.
- Presence of *poor ocular surface, cataract or posterior capsular opacification* could lead to an incomplete scan or one with poor quality. The presence of a lenticular or capsular opacification leads to a false underestimation of the RNFL thickness and a new baseline needs to be established after cataract extraction or capsulotomy (Figure 10.3).
- A patient in the *extremes of age* would not have appropriate normative database and the scan interpretation would be fallacious.
- Patients of *different races* could have potentially different interpretation due to variations in the normative database.

116 Optical Coherence Tomography in Current Glaucoma Practice

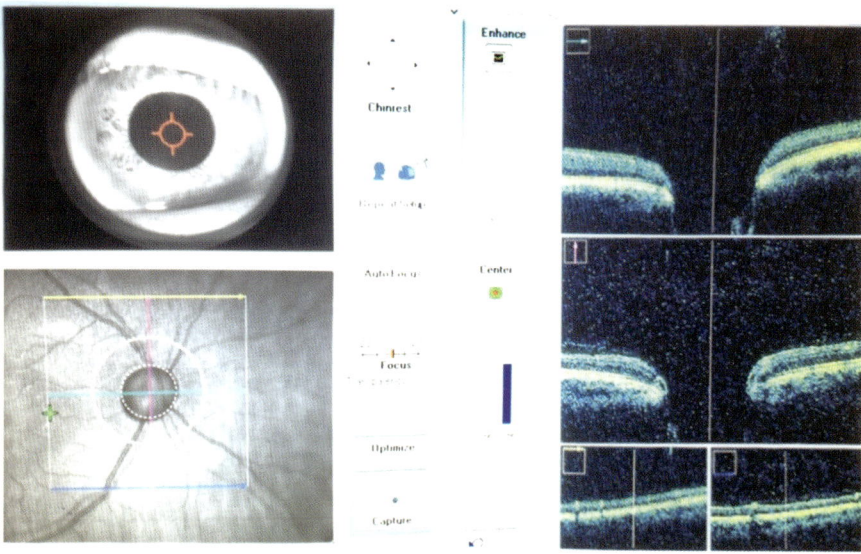

Fig. 10.2 A defocused scan showing the edge of the RNFL and optic disc get cut off during the acquisition process

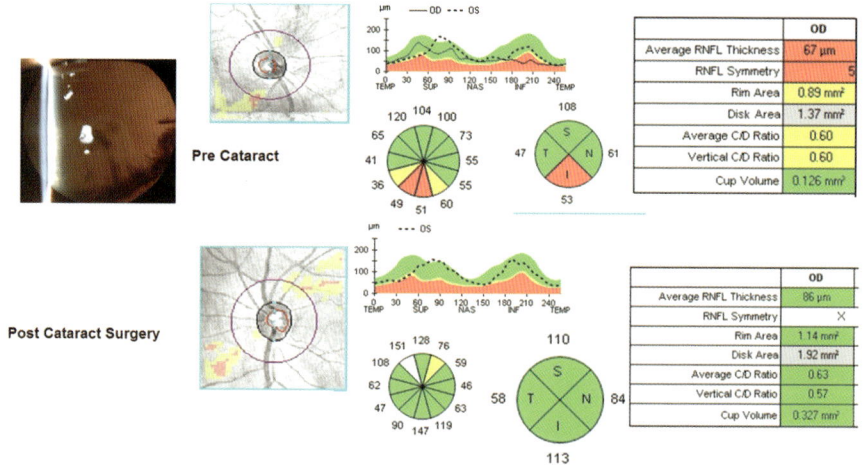

Fig. 10.3 The RNFL OCT of a patient acquired prior to cataract surgery (above) shows RNFL thinning. However after the cataract surgery, the RNFL is seen to be normal (below). This demonstrates the effect of cataract on the OCT analysis

Operator Factors

- *Improper demography:* If the operator enters the age or gender of the patient wrongly, then the interpretation based on age matched normative database would be wrong.[2]
- Image acquisition with the desired area towards the edge of the scan could lead to *edge effect.*
- Operator to *operator variation* is there with regard to focusing and acquisition circle placement.

Ocular Factors

- *Refractive error:* High refractive errors, particularly myopia, tend to show RNFL thinning since (i) The normative database does not include high myopes, (ii) The longer axial length impairs proper focussing, (iii) There may be a posterior staphyloma or excessive peripapillary atrophy.
- *Vitreous floaters:* Presence of shadow artifacts on the scan.
- *Posterior vitreous detachment:* As the PVD tugs on the retina, initially the RNFL may appear to be thickened as it is pulled up by the vitreous but on the completion of PVD, it may suddenly appear thinned due to release from the vitreous face. This may mimic progression and has to be interpreted with caution.
- *Floor effect:* The RNFL thickness can never normally go to zero in a well acquired scan as the glial tissue forms a base and even in the absence of RNFL in an area, it appears to give a reading of 40 to 50 microns. This is called the floor effect.
- *Epiretinal membrane or thick posterior hyaloid:* These may affect the macular scan and lead to a falsely increased thickness of the macula and RNFL.
- *Pigment epithelial defects:* These appear as sharply demarcated areas of RNFL thinning.
- *Nonglaucomatous optic neuropathy:* Optic neuropathy of various causes such as optic neuritis, compressive neuropathy and toxic neuropathy need to be clinically distinguished from glaucoma as they can cause misdiagnosis.

Software Factors

- *Quadrantic averaging:* The quadarantic averaging may miss out localized RNFL defects and hence the OCT may appear normal. A close examination of the clock hour analysis and RNFL profile may disclose such a defect.
- In severe *peripapillary atrophy* of any cause, the ability of the software to detect changes may be reduced. This holds particularly true for progression analysis.

■ HIGH MYOPIA

Case Study

A young high myopic patient was suspected of having juvenile open angle glaucoma. There is a significant cupping and a large disc (Figures 10.4A and B) which shows a thinning of the RNFL on OCT (Figure 10.5) despite a normal visual field (Figures 10.6 and 10.7). However, this thinning is a false red disease and the HRT shows that the neuroretinal rim is essentially normal (Figures 10.8 and 10.9). Thus, this is a case of myopia only and not glaucoma.[2]

Discussion

In order to get maximum benefit from the OCT, it is important to clinically correlate the report and interpret the functional and structural changes together. No single imaging modality is enough to conclude the diagnosis of glaucoma and the cause for a red disease on OCT can be many others than truly glaucoma itself.

Kang et al had studied the "Effect of myopia on the thickness of the retinal nerve fiber layer measured by Cirrus HD optical coherence tomography" and they concluded that the axial length affected the average RNFL thickness, and myopia affected the RNFL thickness distribution. High myopes are likely to exhibit different RNFL distribution patterns. Since ocular magnification significantly affects the RNFL measurement in such patients, it should be considered in diagnosing glaucoma.[3]

It is important in such cases not to start therapy based on RNFL thinning detected on OCT at a single time point, but to do a serial analysis over time (every 6 months-one year) with disc photography to detect for progressive changes in the neuroretinal rim/RNFL before diagnosing glaucomatous optic neuropathy. Eyes with high myopia/pathological myopia more than 12 D are not included in the normative database and imaging in these cases should be interpreted with caution.

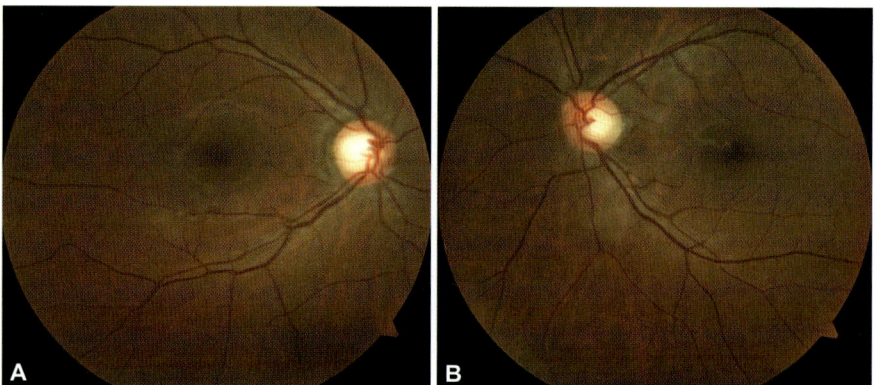

Figs 10.4A and B Fundus image of high myopia patient showing large disc and cups

Limitations of OCT and Imaging Artifacts 119

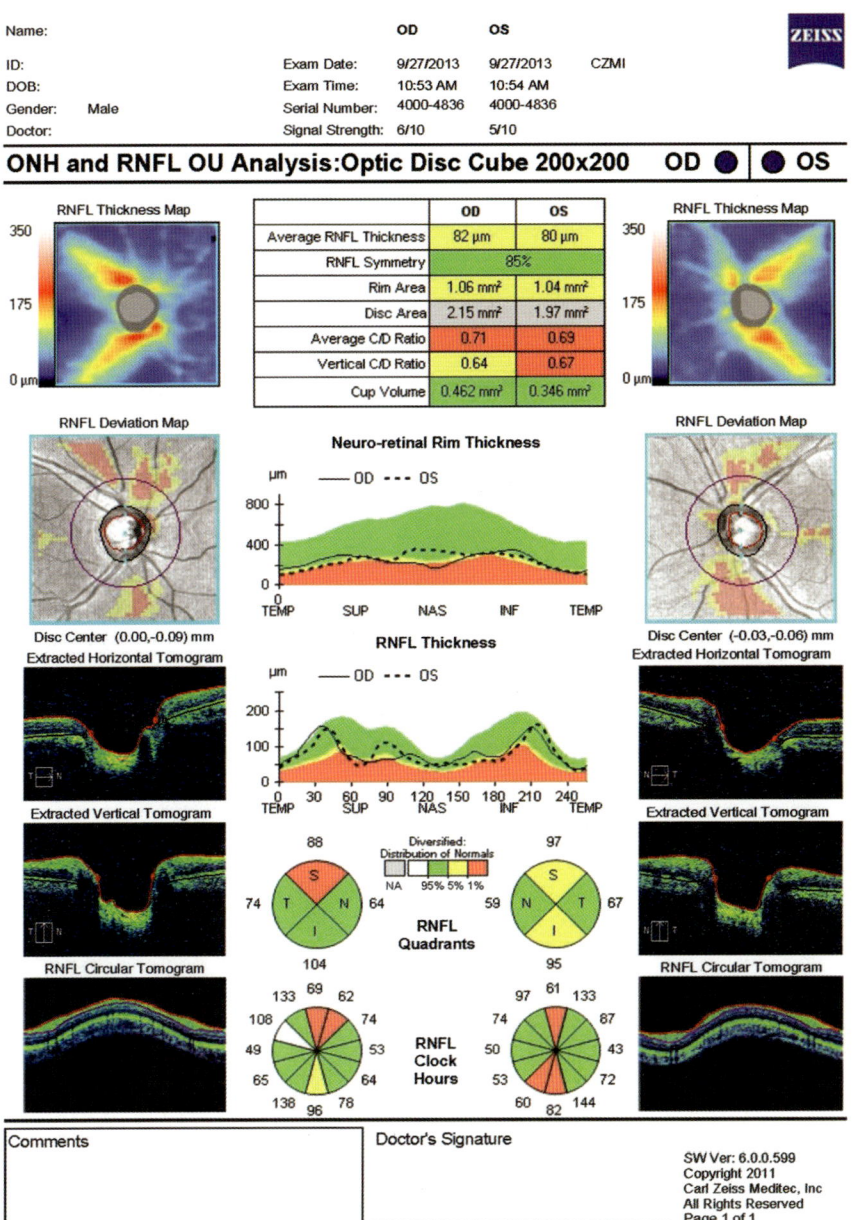

Fig. 10.5 OCT image of high myopia patient showing RNFL thinning in superior quadrant of right eye and superior and inferior quadrant of left eye

120 Optical Coherence Tomography in Current Glaucoma Practice

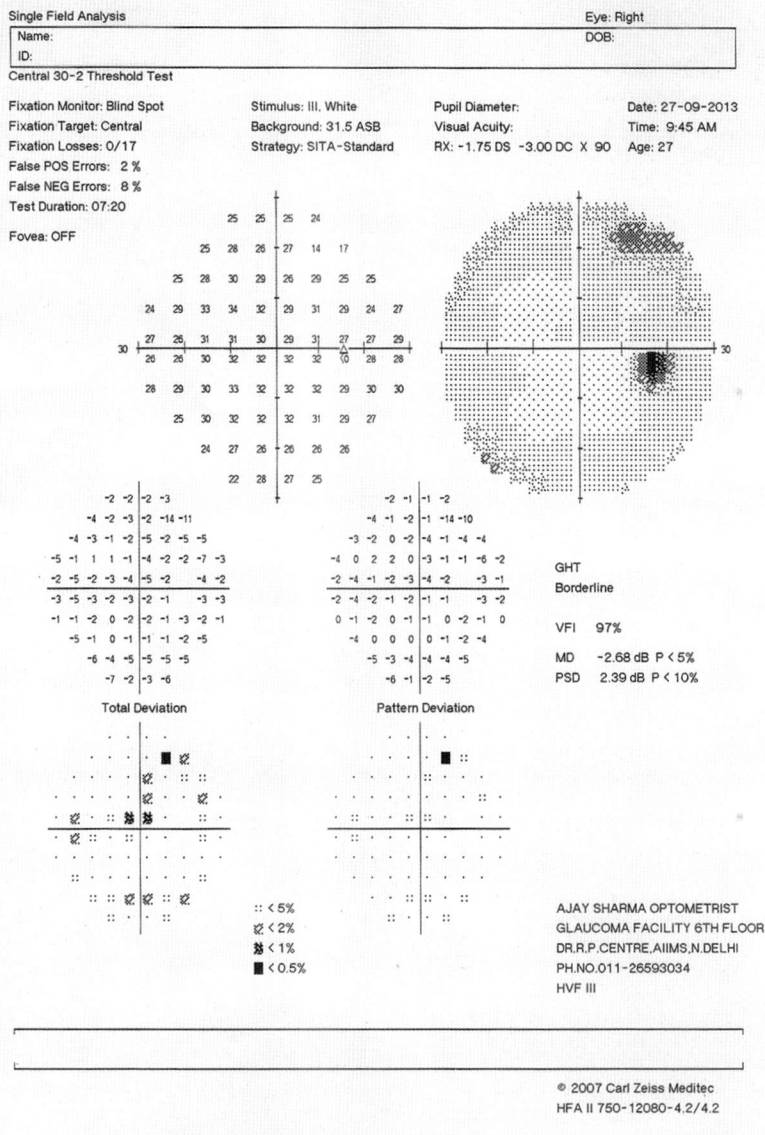

Fig. 10.6 Visual fields of the right eye of high myopia patient do not show any glaucomatous change

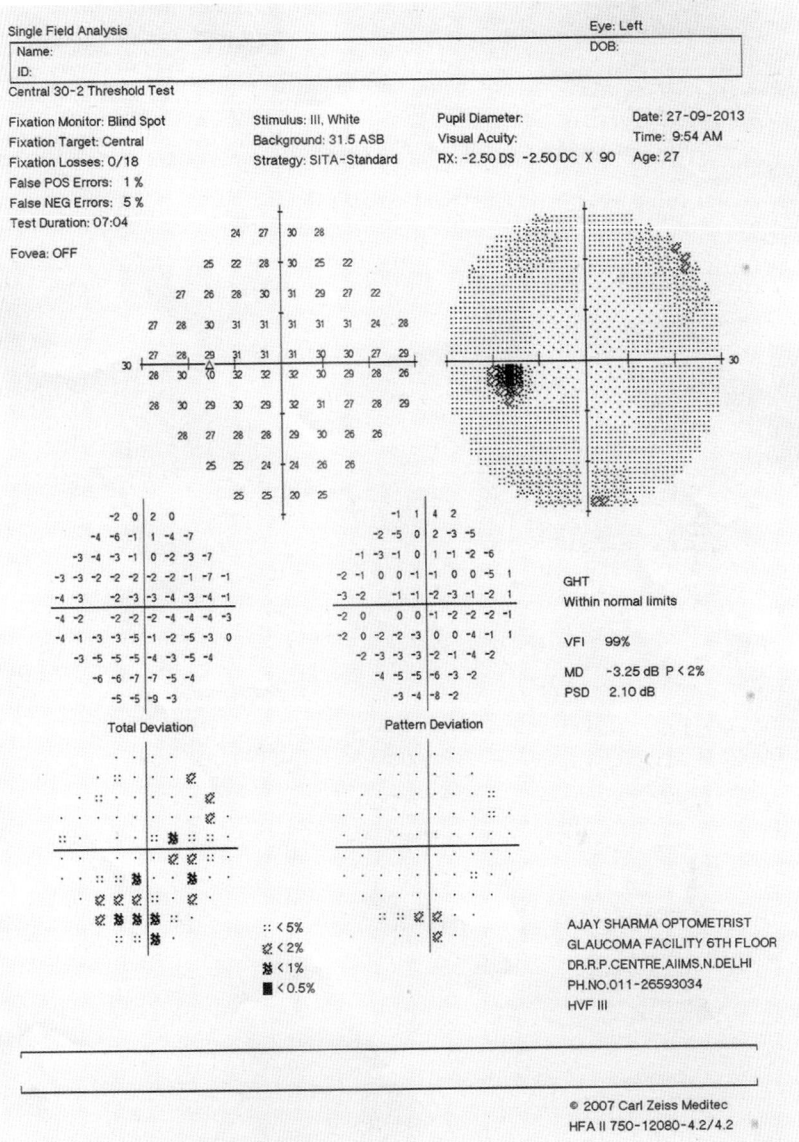

Fig. 10.7 Visual fields of the left eye of high myopia patient do not show any glaucomatous change

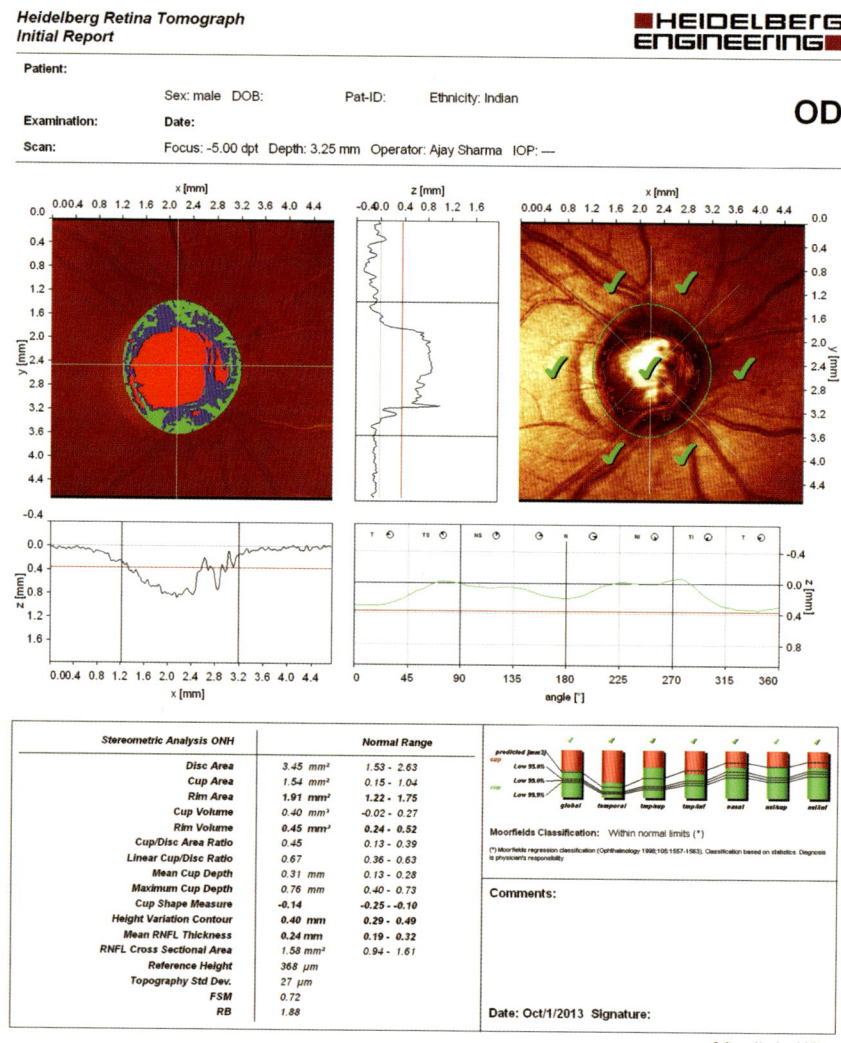

Fig. 10.8 HRT image of high myopia patient showing a healthy neuroretinal rim of right eye

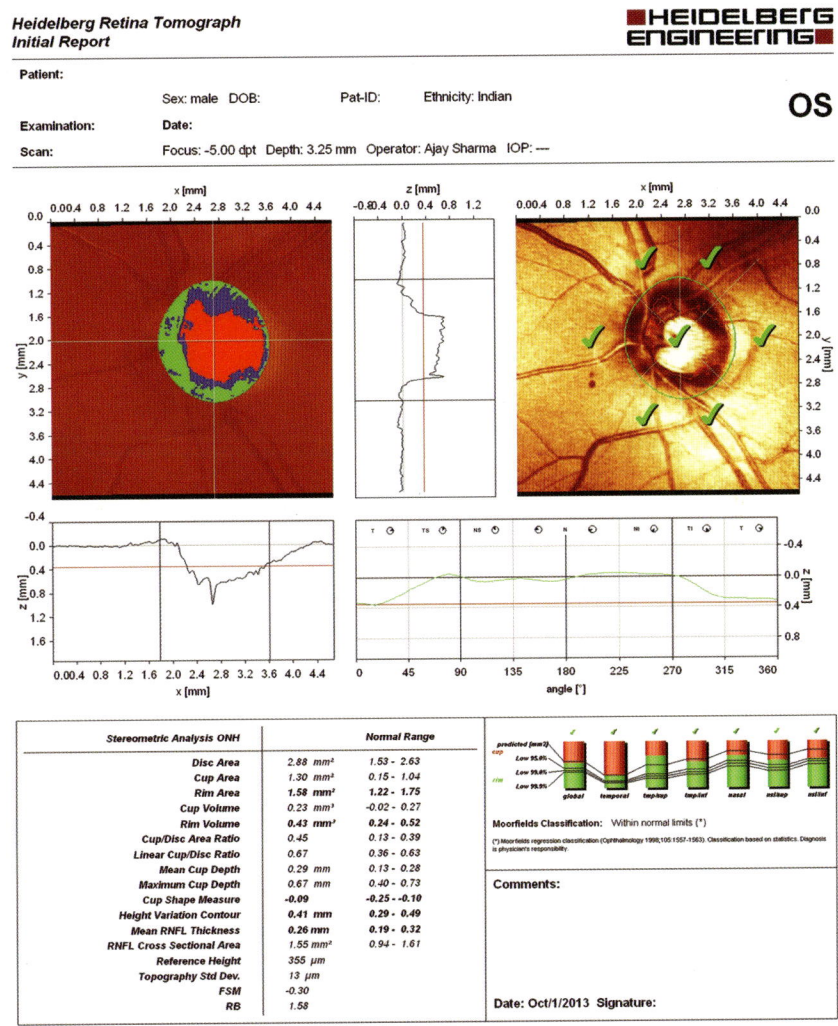

Fig. 10.9 HRT image of high myopia patient showing a healthy neuroretinal rim of left eye

CONCLUSION

In conclusion, glaucoma diagnosis requires documentation of both structure and function. The OCT is an excellent test to document baseline structure of the RNFL, optic nerve head and ganglion cell complex and evaluate progressive changes over time caused by the disease process. If any changes in RNFL are documented on OCT at any time in a patient having manifest/suspected glaucoma always re-confirm by repeat testing ensuring that there is no media opacity, the scan is well centered and the quality of the scan is 7 or more. The follow-up OCT may be done at interval of 6 months in patients with evidence of progression or IOP not at target and at 12 monthly intervals for glaucoma patients who are well controlled/ocular hypertensives/glaucoma suspects.

Never interpret an abnormal OCT in isolation as Glaucoma. Always correlate with clinical findings—look at the RNFL with red free light and the optic nerve head with a stereoscopic examination on the slit lamp with a 78 D or 90 D lens. Always rule out other causes of optic neuropathy other than glaucoma—especially if the optic disc pallor is more than the cupping. Such cases may require neuro-imaging to rule out pathology in the central nervous system (e.g. intracranial tumor, etc.).

At this time point, the use of OCT is not recommended as a screening test for glaucoma. However, it provides useful information to the ophthalmologist who understands its advantages and limitations and can serve as an important aid in the diagnosis and management of glaucoma patients. Always remember that no machine is better than what the human eye and brain can detect.

Key Points

- Acquire a scan with a proper protocol to achieve reproducible and accurate results.
- Various factors including machine factors, operator factors and patient factors can alter the scan results.
- Poor ocular surface, lenticular opacification, posterior vitreous detachment and epiretinal membranes can lead to errors in scan results.
- Poor signal strength, head tilt and improperly calibrated machines can lead to false interpretation of scans and pseudoprogression.
- Always interpret a scan in conjunction with clinical findings including the fundus and visual fields.

REFERENCES

1. Rao HL, Addepalli UK, Yadav RK, Senthil S, Choudhari NS, Garudadri CS. Effect of scan quality on diagnostic accuracy of spectral domain optical coherence tomography in glaucoma. Am J Ophthalmol. 2013 Dec 14.
2. Chong GT, Lee RK. Glaucoma versus red disease: imaging and glaucoma diagnosis. Curr Opin Ophthalmol. 2012;23(2):79-88.
3. Kang SH, Hong SW, Im SK, Lee SH, Ahn MD. Effect of myopia on the thickness of the retina nerve fiber layer measured by Cirrus HD optical coherence tomography. IOVS. August 2010, Vol. 51, No. 8.

Index

Page numbers followed by *f* refer to figure and *t* refer to table

A
Advanced glaucoma 94
Anterior segment
    analysis 20*f*, 106
    cube program 107*f*, 109*f*
    evaluation 17
    examination 13
    OCT 13, 17
    visante OCT 17
Average
    cup-to-disc ratio progression graph 59*f*
    RNFL thickness 39, 58

B
Bulbar conjunctiva of bleb 111*f*

C
Cirrus optical coherence tomography machine 5*f*
Clock hour RNFL thickness maps 42, 42*f*
Correct method of scan acquisition 114*f*
Cup
    area 26
    volume 27
Cystoid macular edema 103*f*

D
Deterioration of optics of OCT 115
Deviation map 89, 90*f*

E
Epiretinal membrane 117
Extracted
    B scan of
        optic nerve head 35
        RNFL 39
    vertical tomogram 39*f*

F
Fast RNFL thickness protocol 35

G
Ganglion cell
    analysis 17, 18*f*, 87, 89*f*, 90*f*, 92
    layer 88*f*
    layer thickness 13
Guided progression analysis 13, 16, 16*f*, 55

H
Healthy neuroretinal rim of
    left eye 123*f*
    right eye 122*f*
Heidelberg retinal tomography 3*t*
    machine 7*f*
High myopia 118
Horizontal
    and vertical B-scans 91
    integrated rim width 27
Hypotonic maculopathy 104*f*

I
Inner plexiform layer 88*f*

L
Lamina cribrosa 13
    cribrosa imaging 20
Laser interferometry 2
Layers of retina 88*f*
Limitations of OCT and imaging artifacts 113

M
Macular
    analysis 13, 19*f*
    cube analysis 101*f*
    OCT 100
Michelson interferometry 9

N
Nasal and inferior quadrants 15
Neuro-retinal rim
    area 26
    thickness 28

Nonglaucomatous optic neuropathy 117
Normal macular thickness 101*f*

O

Optic nerve head
    analysis 13, 14*f*, 24
        scan 26*f*, 27*f*
    gray scale 35
Optical coherence tomography 1, 3*t*, 9, 13, 15, 33, 55, 100
    in glaucoma 13

P

Peripapillary atrophy 117
Pigment epithelial defects 117
Posterior vitreous detachment 117
Preperimetric and perimetric glaucoma 45, 96
Principle of optical coherence tomography 9
    tomography machine 9*f*
Pseudoprogression 68

Q

Quadrantic RNFL thickness maps 42, 42*f*

R

Reconstruction of optic nerve head 29*f*
Refractive error 117
Retinal nerve fiber
    analysis 33
    layer 25, 55, 87
        layer analysis 14
        loss 1
        thickness 100
        thickness map 58*f*
        thickness measurement 13

Retinal pigment epithelium 25
RNFL
    analysis 15*f*, 38*f*
    deviation map 35, 37*f*
    map 35
    profile
        graph 59
        thickness progression graph 59*f*
    thickness 35
        change map 58, 58*f*
        data table 40*f*
        map 38*f*, 39, 40, 58
        progression graphs 59*f*
        protocol 34

S

Scanning laser
    ophthalmoscopy 2
    polarimetry 2
Severe peripapillary atrophy 117
Stratus optical coherence tomography machine 4*f*

T

Thick posterior hyaloid 117
Thickness
    acquisition map 88, 90
    table 89, 90
Time domain and spectral domain OCT 11*t*
TSNIT
    neuroretinal rim thickness profile 40, 41*f*
    RNFL thickness profile 40, 41*f*

V

Visante OCT anterior segment analysis 22*f*
Vitreous floaters 117